NETWORKED NATION: BROADBAND IN AMERICA 2007

NETWORKED NATION: BROADBAND IN AMERICA 2007

PETER ACKERMAN
AND
OLIVER MABEL
EDITORS

Nova Science Publishers, Inc.
New York

For permission to use material from this book please contact us:
Telephone 631-231-7269; Fax 631-231-8175
Web Site: http://www.novapublishers.com

LIBRARY OF CONGRESS CATALOGING-IN-PUBLICATION DATA
Available upon request
ISBN: 978-1-60692-811-0

Published by Nova Science Publishers, Inc. ✢ New York

CONTENTS

PREFACE[*]

Broadband technology is changing our lives, our economy, and our culture. By making it possible to access, use, and share information, news, and entertainment with ever increasing speed, broadband knits geographically-distant individuals and businesses more closely together, increases our productivity, and enriches our quality of life. In so doing, it fuels economic growth and job creation that, in turn, provide unparalleled new opportunities for our nation's citizens.The procompetition telecommunications policies underpinning the administration's broadband strategy is examined in this book. Among other things, it stabilized the market for planning and investment, provided needed spectrum resources to expand existing services and support new innovations, and spurred the development and deployment of technologies and infrastructure necessary to deliver them. The same technology that enables governments, banks, corporations and other institutions to manage affairs of state and international finance and trade now makes critical contributions to health care, education, public safety, and the productivity of individuals and small businesses, as well as a host of other activities. The swift growth in broadband technologies and services that has occurred over the last several years are also discussed in this book. This is an edited and excerpted edition of a National Telecommunications and Information Administration publication.

---

[*] Excerpted from National Telecommunications and Information Administration, dated January 2008.

EXECUTIVE SUMMARY

Broadband technology is changing our lives, our economy, and our culture. By making it possible to access, use, and share information, news, and entertainment with ever increasing speed, broadband knits geographically-distant individuals and businesses more closely together, increases our productivity, and enriches our quality of life. In so doing, it fuels economic growth and job creation that, in turn, provide unparalleled new opportunities for our nation's citizens.

Recognizing this transformative power, four years ago President Bush articulated a national vision: universal, affordable access to broadband technology. From its first days, the Administration has implemented a comprehensive and integrated package of technology, regulatory, and fiscal policies designed to lower barriers and create an environment in which broadband innovation and competition can flourish.

The results have been striking. The last several years have witnessed substantial growth in the broadband marketplace punctuated by increases in capital investment, innovation, and market entry. Relative to other countries, the United States has experienced superior productivity over the past several years. Americans today enjoy an increasing array of broadband services, available from a growing number of service providers, using a variety of technologies. Penetration continues to grow, and prices continue to fall.

A Broadband Strategy for the Twenty- first Century

Building on the pro-competitive, deregulatory policies Congress articulated in the Telecommunications Act of 1996, the Administration, through the combined efforts of the National Telecommunications and Information Administration (NTIA) of the U.S. Department of Commerce, the U.S. Department of Agriculture (USDA), and other agencies, and in cooperation with the Federal Communications Commission (FCC or Commission), has executed a combination of initiatives to develop and rapidly deploy new technologies, eliminate regulatory underbrush, and remove economic disincentives for investment in this critical area.

Technology Policies

Technology Neutrality: Past experience teaches that when government tries to substitute its judgment for that of the market by favoring one product or vendor over another, it can easily divert investment and/or discourage research necessary to bring new and better products or services to market. Given the rapid pace of technological change, such unintended effects can have longterm and far-reaching adverse consequences that extend across multiple sectors of the economy. For this reason, the Administration has consistently and strenuously advocated for technology neutrality in order to take the government out of decisions more appropriately left to the marketplace.

Spectrum Policy

To satisfy society's increasing desire to function "untethered," new wireless broadband services can only emerge if spectrum resources exist to support them. Pursuant to the President's June 2003 Spectrum Policy Initiative, NTIA has promoted greater access to spectrum by improving the spectrum efficiency of Federal radio systems and operations. NTIA and the FCC, working together and with the Congress, have significantly increased the amount of spectrum available for advanced wireless services. The Administration has also actively sought ways to accelerate broadband deployment by facilitating unlicensed uses and encouraging the use of promising new spectrum technologies such as ultrawideband, and by implementing new spectrum management systems that afford non- Federal users faster and easier access to spectrum.

Regulatory Policies

The Administration has worked to clear away regulatory obstacles that could thwart the investment that fuels development – and deployment – of new technologies. It supported the FCC's efforts to modify legacy regulation in order to expand incentives for local telephone companies to invest in network upgrades and to stimulate facilities-based investments by other providers. The Administration has further encouraged aggressive broadband deployment by favoring policies to ensure that applications running on the Internet, such as voice over Internet Protocol service, remain free from unnecessary economic regulation. The Administration has also supported cable franchise reform efforts in order to promote the video services competition that will help to accelerate broadband deployment. In April 2004, the President signed an Executive Memorandum to give broadband providers more timely and cost-effective access to rights-of-way on Federal lands for their networks.

Fiscal Policies and Targeted Funding Efforts

Tax relief has given businesses powerful incentives to invest in broadband technology. Recognizing that taxes constrain growth, President Bush on three occasions has signed legislation to extend the Federal moratorium on State and local taxation of Internet access and has consistently urged that the moratorium be made permanent. The President also signed into law legislation allowing companies to accelerate depreciation for capital expenditures, including those associated with broadband deployment. The Administration has also forcefully advocated extending and making permanent the tax credit for research and development spending. Further, the Administration has provided targeted seed-funding to support more rapid deployment of broadband in underserved rural areas, in particular through several programs administered by the USDA.

PROGRESS IN BROADBAND ACCESS

America's consumers are now reaping the rewards of the Administration's pro-investment, deregulatory policies: a vigorous broadband marketplace in which providers using various platforms compete against one another on price, speed, mobility, content, and other service features. Currently available data suggest that broadband availability and subscribership have increased dramatically, and that consumers – including those in rural and remote areas – have more opportunities than ever to choose the broadband solution (*i.e.*, technology, services, and provider) that best suits their needs and budget.

Overall

Since President Bush took office, the total number of broadband lines in the United States has grown by over 1,100 percent from almost 6.8 million lines in December 2000, to 82.5 million in December 2006 according to the most recent FCC data. Over 58 million of these lines serve residential customers. FCC data also reveal that the number of broadband service providers more than tripled from December 2003 to December 2006, with the newest wired and wireless services growing at the highest rates. Data from the U.S. Census Bureau's Internet Use Supplement to the October 2007 Current Population Survey also show that Americans' use of broadband technologies has soared: home broadband usage has risen substantially over the past six years, from 9.1 percent of households in September 2001 to 50.8 percent in October 2007. Rural America has also experienced impressive growth from 5.6 percent of rural households in September 2001 to 38.8 percent in October 2007.

Digital Subscriber Line (DSL)

Used primarily by local telephone companies to provide broadband services, DSL was available as of year end 2006 to 79 percent of households in areas where companies also offered local telephone service according to FCC data. Not limited to highly populated areas, DSL was among the broadband services offered to 90 percent of the carriers' rural customer base according to recent industry surveys. As DSL transmission speeds have accelerated and prices have dropped, the number of DSL subscribers has burgeoned. FCC's statistics reveal about 22.8 million residential "high-speed" asymmetric DSL (ADSL) lines in service as of year end 2006, growing by more than 300 percent from December 2002.

High-Speed Cable

Research conducted by the cable television industry suggests that high-speed cable service is now available to some 92 percent of all U.S. households by the end of 2007. That same research shows that as of year end 2006, there were 29 million residential high-speed cable subscribers, up more than 109 percent from the 13.8 million customers in the second quarter of 2003. Accompanying this growth, the industry's trade association reports that most cable broadband operators now offer transmission speeds exceeding five megabits per second (5 Mbps), with some delivering as much as 50 Mbps.

Mobile Wireless

The wireless industry is currently the fastest growing segment of America's broadband economy. The FCC's most recent data reveal that the number of broadband lines provided by wireless operators increased from approximately 380,000 in June 2005 to almost 22 million at year end 2006 – a growth rate that dwarfs that of other broadband platforms. Moreover, a significant portion of these (4.1 million, or almost 19 percent) provide "advanced" services capable of delivering over 200 kilobits per second (Kbps) in both directions.

Fixed Wireless

Fixed wireless technologies have emerged both as an important complement to mobile wireless and as a potential "last-mile" broadband solution in areas that other platforms do not reach. According to FCC figures, the number of fixed wireless broadband lines in the United States grew 132 percent from 208,695 in June 2005 to 484,073 in December 2006; an industry group estimated the total number of fixed wireless subscribers in 2006 to be 800,000. The proliferation of one fixed wireless technology – wireless fidelity (Wi-Fi) – has increasingly enabled many consumers to cut the tether to a home or office connection; based on one estimate, there are 66,058 public/commercial Wi-Fi access points, or "hot spots", currently in the United States, more than double the next closest country. Another technology – worldwide interoperability for microwave access (WiMAX) – can deliver fixed wireless broadband access at distances as great as five miles without line of sight and up to 30 miles under ideal conditions. With potential data speeds up to 70 Mbps, WiMAX has been identified as a possible "last-mile" solution to deliver broadband into currently unserved rural and remote areas.

Satellite

With coverage of virtually the entire continent, satellite has become the default solution for areas not covered by terrestrial sources of broadband. Like those other broadband services, satellite broadband has seen tremendous growth over the past few years. From fewer than 50,000 subscribers in 2004, satellite providers were serving an estimated 700,000 subscribers at year end 2006.

Fiber Optic and Broadband over Power Lines

FCC data show that the total number of high speed lines delivered over fiber and power line connections grew 789 percent from December 2003 to December 2006, rising from 116,390 to just over a million. Fiber optic lines, however, appear to be almost entirely responsible for this expansion. Industry data also show a steady expansion of fiber deployment (including fiber to the home and interoffice fiber). As charted by the Telecommunications Industry Association (TIA), annual deployment of fiber in the United States grew sharply over the last four years, from the 4.8 million miles deployed in 2003 to 13.1 million miles in 2007. According to another industry estimate, fiber now passes almost 9.6 million homes in North America (virtually all of these homes are in the United States), up more than 50 percent from September 2006.

Industry Investment

The accessibility of broadband technology to an increasing number of Americans stems directly from the substantial and growing capital investments made by service providers across all of the subsectors of the broadband marketplace. TIA estimates that overall spending to support broadband network infrastructure will rise over the next four years from $15.2 billion in 2007 to $23 billion in 2010. This includes: significant growth in fiber deployment by cable operators to meet customer demand for high-definition television channels and video-on-demand services; continued investment for the deployment of advanced mobile wireless services and fixed wireless technologies such as WiMAX; and spending on fiber-related telecommunications equipment that is estimated to grow almost $13.5 billion annually in the next three years.

Affordability

Escalating competition among broadband platforms and service providers has yielded both a proliferation of new communications and entertainment services and affordable broadband pricing for American consumers. The rapid drop in the prices for DSL broadband services in the past five years, along with the deployment of fiber infrastructure equipment, have resulted in lower prices for high-speed cable services. Despite the dramatic increase in usage, the average local monthly bill for wireless services has remained relatively stable, rising only a dime per year on average over the past four years. Satellite broadband prices have also dropped as satellite providers introduce alternative pricing models to make their services more affordable.

Chapter 1

INTRODUCTION

- *A surgeon in Queensland, Australia, using an Internet-based laser technology, RoboLase, performs microsurgery successfully on cells located in a laboratory in San Diego… .* [1]
- *From over 6,000 miles away, an American soldier stationed in Iraq watches via live web- cast as his 18-year-old daughter in Tennessee, his oldest child, walks across the stage to accept her high school diploma… .* [2]
- *With only a laptop and a wireless card, a farmer sitting in his onion field in Hermiston, Oregon, can check e-mail, monitor the water level of his crops, and even conduct business with his customers, all using the 700 square mile Wi-Fi "cloud" that covers his farm … .* [3]
- *Using a national, interoperable communications service, public safety personnel and first responders in Dallas can communicate with one another with voice, data, and real-time live video feeds to maintain situational awareness and coordinate response activities more effectively … .* [4]
- *Students in Kotzebue, Alaska (66.53N, 162.39W), on Alaska's Western shore facing the Bering Strait, join with students in other remote communities on Alaska's Southwestern frontier, in a live "virtual field trip" to the Johnson Space Center in Houston and two weeks later visit with an Alaskan astronaut aboard the space shuttle Discovery, all using a state broadband Distance Learning network.* [5]

The foregoing examples illustrate only a few of the ways in which broadband technology is changing our lives, our economy, and our culture. By making it possible to access, use, and share information, news, and entertainment with ever increasing speed, broadband technology knits geographically-distant individuals and businesses more closely together, increases productivity, and enriches the quality of life. In so doing, it catalyzes economic growth and job creation that, in turn, provide unparalleled new opportunities for our nation's citizens.

Recognizing this transformative power and realizing the benefits that broadband technologies can deliver for all Americans, President Bush four years ago articulated a national vision to carry the United States forward: universal, affordable broadband access. Even before advancing that goal, however, the Administration had started to pursue it, implementing a comprehensive and integrated package of technology, regulatory, and fiscal policies designed to lower barriers to competition and investment, and create an environment in which broadband innovation and competition could flourish.

The results have been striking. The last several years have witnessed substantial growth in the broadband marketplace punctuated by demonstrable increases in capital investment, innovation, and entry, as well as superior productivity relative to other countries. Americans today enjoy an increasing array of broadband services available from a growing number of service providers across multiple platforms – traditional wireline, cable, fiber, wireless, satellite, Wi-Fi and WiMAX. Penetration rates continue to grow, and prices continue to fall.

This Report examines the concerted, pro- market policies that the Administration has implemented to remove regulatory obstacles, foster economic incentives, and encourage investment in the development and deployment of broadband technologies. It also outlines steps the Administration has taken to help jump-start market forces by providing seed-funding to advance broadband deployment in underserved and rural areas. Finally, it documents the statistical evidence that demonstrates the burgeoning competitive broadband marketplace that has emerged in the United States over the last several years and the commensurate increases in accessibility and affordabili ty that American consumers enjoy as a result.

Chapter 2

A BROADBAND STRATEGY FOR THE TWENTY-FIRST CENTURY

The Administration's strategy to advance broadband infrastructure and services in the United States rests on a simple tenet: A competitive environment – in which the ingenuity of technologists, engineers, and businessmen can fully express itself – provides the most effective and reliable tool to identify the new technologies, products and services that consumers need and want; and to develop and deploy the most cost-effective technical and commercial solutions to fill those needs. This precept is not new: it is the enduring legacy of a strong, bi-partisan consensus forged over a decade ago.

In enacting the Telecommunications Act of 1996, overwhelming majorities in both chambers of Congress and the Executive Branch enshrined the policy "[t]o promote competition and reduce regulation in order to secure lower prices and higher quality services for American telecommunications consumers and encourage the rapid deployment of new telecommunications technologies."[6] Mindful of the remarkable potential of high-speed, high- capacity, digital communications systems, the Act specifically embraced as "the policy of the United States… ."

> to promote the continued development of the Internet and other interactive computer services and other interactive media; …. [and] preserve the vibrant and competitive free market that presently exists for the Internet and other interactive computer services, unfettered by Federal or State regulation… .[7]

Building on the policies Congress articulated in the Telecommunications Act, the Administration has maintained a principled approach to telecommunications policies, recognizing that:

> "The role of government is not to create wealth; the role of our government is to create an environment in which the entrepreneur can flourish, in which minds can expand, in which technologies can reach new frontiers."
>
> President George W. Bush, Technology Agenda, June, 2002

- New technologies and services, such as those supported by Internet Protocol (IP)-based networks, are creating competition in the telecommunications market, allowing for deregulation;
- New technologies are making distance irrelevant – and because the communications services supported by these technologies are inherently interstate in nature, jurisdiction over them should reside at the Federal, rather than the State, level;
- Specific social policy objectives – such as support for 911 services, accessibility for persons with disabilities or those residing in rural or remote areas, and the needs of law enforcement – may require regulation even in a competitive environment; and
- Telecommunications subsidies should reflect the economics of new technologies and actual consumer needs.

These principles serve as the foundation for three interlocking, mutually reinforcing elements that form the President's broadband strategy – technology policy, regulatory policy, and fiscal/economic policy. The Administration, through the efforts of the National Telecommunications and Information Administration (NTIA) of the U.S. Department of Commerce, the U.S. Department of Agriculture (USDA), and other agencies, and in cooperation with the Federal Communications Commission (FCC or Commission), has executed a combination of initiatives to hasten the development and deployment of new technologies, eliminate regulatory underbrush, and remove economic disincentives for investment in this critical area. Some of the specific initiatives in each of these areas are discussed more fully below.

TECHNOLOGY POLICIES

From the outset, fostering innovation and technical invention as an engine for U.S. economic growth has been a core priority of this Administration. Consequently, promoting the development and deployment of broadband networks has been a centerpiece of the President's technology policies. In June 2002, the President released his Technology Agenda.[8] In it, he observed that "[n]ew technologies are revolutionizing the way we learn, work and play – all while improving our quality of life," and that "high speed data networks known as broadband offer great potential to increase productivity, promote economic growth and revolutionize how we deliver health care."[9]

The Technology Agenda recognizes that broadband technology draws momentum from pro-technology policies, amplifies it, and transfers that force to other sectors of the economy to stimulate innovation, competitiveness, and overall growth. To accelerate that momentum, the President advanced a number of initiatives including strengthening research and development funding; education reform, especially in the areas of math and science; and support for entrepreneurship. For example, to stimulate innovation, the Administration championed increases in Federal research and development funding, elevating it to almost $143 billion, an increase of 56 percent above 2001 levels and the largest Federal research and development budget in history.

Central to this effort is the President's American Competitiveness Initiative, which seeks to double investment in key civilian Federal science agencies (the Department of Commerce's National Institute of Standards and Technology, the Department of Energy's Office of Science, and the National Science Foundation) that support innovation-enabling basic research in physical sciences and engineering.[10] Over three billion dollars was budgeted for the National Information Technology Research and Development (NITRD) program, which includes research directly related to broadband technology.[11] The Technology Agenda also specifically charged the President's Council of Advisors on Science and Technology (PCAST) to examine consumer-demand issues and provide recommendations for steps to facilitate broadband deployment.[12]

In addition to the general measures identified above, the Administration also implemented a number of targeted pro-growth telecommunications policies that have also contributed to robust technological development in the broadband sector. These include continuing support for technological neutrality, efforts to unlock the economic value and entrepreneurial potential of U.S. spectrum assets, and removing barriers to innovative new applications such as ultrawideband technology and broadband over power lines (BPL).

Recommendations of the President's Council of Advisors on Science and Technology (PCAST): TELEWORK

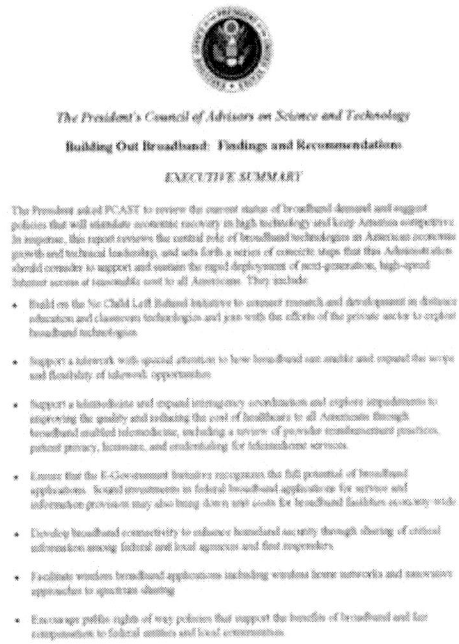
The President's Council of Advisors on Science and Technology

Building Out Broadband: Findings and Recommendations

EXECUTIVE SUMMARY

In its report to the President, the PCAST identified several measures that it believed could contribute to more rapid deployment of next-generation, high-speed systems. Among them, the panel specifically urged support for telework, "with special attention to how broadband can enable and ex-pand the scope and flexibility of telework opportunities." Report on Building Out Broadband, President's Council of Advisors on Science and Tech-nology at 3 (2002), http://www.ostp.gov/PCAST/FINAL%20Broadband%20Report%20With%20Letters.pdf.

In this regard, a February 2007 Report by WorldatWork (based on re-search by the Dieringer Research Group) found "a 25-percent one-year increase [from 2005 to 2006], and a 63-percent two-year increase" in the number of Americans whose employer permits them to work from home at least one day a month. The Report concluded that this rising trend "is likely [due to] a combination of factors, including the proliferation of high speed/broadband and other wireless access (which has made it less expen-sive and more productive to work remotely) ... ." Telework Trendlines for 2006 – 2007 Survey Brief, WorldatWork (Feb. 2007), http://www.workingfromanywhere.org/news/Trendlines_2006.pdf.

With respect specifically to broadband use by teleworkers, the WorldatWork study found that "[t]he number of teleworkers using a broadband connection at home increased by more than 45 percent in the 2006 survey, fol-lowing an even-larger 65 percent rise in the previous survey (2005)." The Report found that the number of home-based employed teleworkers using broadband rose from 8 million in 2004 to 19.2 million in 2006. It stated that such increases "are logical given the explosion in broadband and high-speed Internet usage in the past several years" which, it observed, "has helped employees more productively work from a distance, especially in accessing corporate networks."

Technology Neutrality

A hallmark of the competitive system embraced in the 1996 Telecommunications Act is consumer choice: the consumer – not government – determines what products and services are needed, the acceptable price for those products or services (*i.e.*, what consumers are willing to pay for them), and which vendor best satisfies the consumer's service needs. Experience teaches that when government tries to substitute its judgment for that of the free market, or otherwise anticipate consumer demand by favoring one product or vendor over another, it can easily distort the marketplace, resulting in the diversion of investment and/or discouraging the research and innovation necessary to bring new and better products or services to market.

In a rapidly evolving area such as information and communication s technology (ICT), such unintended effects can carry fundamental, long-term adverse consequences that extend across multiple sectors of the economy. Accordingly, especially in the area of ICT, the government should exercise care not to dictate standards, but instead should seek to foster an environment in which the potential of competitive forces can be fully realized to maximize the choices (*e.g.*, technologies, products, services, applications, providers) available to consumers. For this reason, the Administration has consistently and strenuously advocated for technology neutrality in order to remove the government from what are more appropriately marketplace decisions.

Domestically, this policy has yielded an array of competing broadband services offered over an equally diverse variety of platforms – wireless, wireline, cable, fiber, satellite, and power line. Consumers need no longer buy discrete services from monopoly providers but can choose from among multiple service providers based on price, performance, mobility, and other features and characteristics. Internationally, the Administration has expressed great concern whenever countries prescribe a standard for technology that impedes competition, obstructs investment, or hampers the creation of competitive markets. The United States has actively worked to encourage all nations to open their markets, to implement measures to protect intellectual property, and to use widely adopted international standards. The Administration has also included technology neutrality principles in U.S. free trade agreements.

Spectrum Policy

Recognizing society's increasing desire to function "untethered," whether for convenience or improved effectiveness, mobility stands out more and more as a critical feature of broadband applications. For example, achieving broadband access at a meeting facility enables business travelers to maintain connectivity with the office. Distributing broadband access via wireless

technology to hundreds of meetings attendees simultaneously multiplies its utility and increases the productivity of the work force. Such applications require spectrum access that is fast, reliable and convenient.

While ensuring a level playing field for all technological platforms, the Federal government – as custodian and manager for a significant amount of our national spectrum assets – can play a major role in technological advancement by making spectrum resources available to engineers and entrepreneurs to support development and implementation of new technologies and services, including broadband communications. The Administration has implemented a number of policies to open up this limited resource by improving spectrum management policies and practices to increase spectrum efficiency, reallocating Federal spectrum so that it can be used for new applications, and maximizing opportunities for the unlicensed use of spectrum wherever possible.

Improving Spectrum Management

In June 2003, the President issued a memorandum to the heads of Executive Departments and Agencies outlining his "Spectrum Policy for the 21st Century."[13] The President's Spectrum Policy Initiative called for a critical review of current spectrum management policies and practices to establish a new U.S. spectrum policy that will foster economic growth; ensure U.S. national and homeland security; maintain our global leadership in communications technology and services; and satisfy other vital U.S. needs in areas such as public safety, scientific research, Federal transportation infrastructure, and law enforcement.

Pursuant to this initiative, NTIA has implemented measures to promote greater access to the radio spectrum by improving the efficiency of radio systems and operations. Specifically, NTIA has (1) modernized Federal spectrum management processes with advanced information technology, aiming toward fully-automated web-based analysis, coordination, and record-keeping capabilities; (2) enhanced spectrum engineering and analytical tools by developing best practices recognized by government and industry; and (3) promoted the use of market-based economic mechanisms in spectrum management by recognizing the role that spectrum value must play in Federal procurement decision-making. Improved and more efficient spectrum management processes provide a firm foundation for an environment that will continue to advance new broadband networks in the future.[14]

Reallocating Spectrum for New Uses

In the past decade, wireless technologies have proliferated at an astonishing pace across the globe. Although the telecommunications sector constitutes only approximately one percent of our nation's economy, experts estimate that it generates up to 10 percent of U.S. economic growth,[15] and wireless services occupy an increasingly important place in that sector. High-Speed Downlink Packet Access (HSPDA) is accelerating cellular data speeds to multiple megabits per second.[16] Mobile phone content offerings supported by such speeds – including music, gaming and video – are coming to market at a breathtaking pace, with revenues expected to rise from $5.2 billion in 2004 to $43 billion worldwide by 2010.[17] Wireless competition in the United States has resulted in a vibrant marketplace. The U.S. wireless communications services revenues are projected to be more than $141 billion in 2007 and are growing approximately 10 percent each year.[18]

Such new services c an only emerge, however, if the spectrum resources exist to support them. Consequently, the Administration has worked to ensure that commercial wireless services have spectrum to compete with incumbents and provide new services to consumers, while at the same time preserving spectrum access for critical Federal systems and public safety services. NTIA and the FCC, working together and with the Congress, have significantly increased the amount of radio spectrum available for advanced wireless services, including wireless broadband, and other services.

For example, the Administration supported enactment of the Commercial Spectrum Enhancement Act[19] (CSEA), which facilitated the reallocation of 45 megahertz (MHz) of spectrum at 1710-1755 MHz from Federal use to commercial use. The Commission reallocated an additional 45 MHz at 2110- 2155 MHz. The resulting 90 MHz were allocated for advanced wireless services including third generation (3G) wireless broadband services and auctioned in the summer of 2006.[20] The $13.7 billion in auction revenues reflects a significant investment in U.S. wireless broadband deployment, which will further contribute to the accessibility and affordability of such services.[21]

The digital television transition in February 2009 will also free up spectrum in the 700 MHz band. The Administration supported adoption of a firm February 17, 2009, date for completion of the transition from analog to digital television broadcasting in order to recover 108 MHz of analog broadcast spectrum for broadband services to support critical public safety and commercial needs.[22]

The Administration has also taken steps internationally to promote the development of advanced wireless services. (*See* inset box.)

Maximizing Spectrum Sharing

In addition to freeing up spectrum for licensed use, the Administration has also actively sought ways to accelerate broadband deployment by facilitating unlicensed uses in bands with Federal operations. For example, NTIA led efforts to allow unlicensed users to share spectrum with military radar operations in 255 MHz in the 5 GHz band.[23] These efforts supported the development of new technologies that employ innovative spectrum sensing and avoidance techniques capable of preventing radio interference to the radar operations. This has resulted in unlicensed mobile broadband Wi-Fi devices with higher data speeds.

NTIA has also performed tests and analysis supporting the use of promising new ultrawideband (UWB) spectrum technology.[24] The agency continues to work with the FCC to review and modify UWB rules to permit innovative technologies to operate while also protecting Federal operations.

In February 2005, NTIA and the FCC launched an online registration site for highspeed wireless links sharing spectrum in the 70-80-90 GHz bands.[25] Since the website became operational, assignments submitted using the website have increased substantially each year. This automated coordination illustrates how information technology and engineering analysis capabilities can be leveraged to afford non-Federal users faster and easier access to spectrum.

Cooperative Testing and Support for New Technologies

The Administration has also studied new technologies and conducted scientific testing to facilitate the use of existing technology platforms and spectrum to deliver innovative new broadband applications. For example, the FCC and NTIA have worked cooperatively to promote the development of Broadband over Power Line (BPL) technology, a potential “third broadband wire into the home,” while safeguarding existing licensed radio services against harmful interference.[26] NTIA conducted a series of measurements and computer simulation that demonstrated that BPL should not pose an interference problem for existing Federal radiocommunication systems.[27]

U.S. INTERNATIONAL AGENDA TO CLEAR THE PATH FOR FUTURE BROADBAND GROWTH

- The Administration's technology initiatives to foster broadband deployment do not stop at our borders. Broadband occupies an important place on our international telecommunications policy agenda as well. Recently, at the 2007 World Radiocommunication Conference in Geneva, Switzerland (WRC07), the United States charted two important victories in its ongoing efforts to ensure global access to spectrum for broadband wireless technologies.
- First, the United States succeeded in having the WRC identify the 700 MHz band for use by advanced wireless systems such as International Mobile Telecommunications 2000 or "IMT-2000" on a regional basis expanding the potential market for wireless broadband offered in the band from the United States to the Americas and the largest economies in Asia. The WRC's actions to harmonize frequency arrangements should, over time, create a marketplace of over 3 billion people for wireless broadband operating in the 700 MHz band. The economies of scale created by the WRC's action, coupled with the 2007 Radiocommunication Assembly action to include WiMAX in the IMT-2000 family of technologies, should result in a larger, more competitive and open market ultimately benefiting U.S. consumers by reducing prices and increasing the availability of advanced wireless services. ITU, *infra* note 143.
- The United States also secured international agreement regarding protection of terrestrial systems in the 2500-2690 MHz band from satellite interference. The 2500-2690 MHz band is globally harmonized for the provision of advanced terrestrial wireless services and is one of the key bands for WiMAX deployment around the world. The United States, like many countries, is implementing WiMAX in the band and will be releasing products and services in the near future to meet consumer demands for broadband wireless access. Interference from satellite systems could have impeded introduction of terrestrial services in the United States, including WiMAX. WRC-07 agreed immediately to place transmission power limits on new satellite systems in order to protect the U.S. terrestrial environment, ensuring stability to satellite-terrestrial spectrum sharing in the 2.5 GHz band and the continued growth of the U.S. broadband wireless market. ITU, *Provisional Final Acts of WRC-07*, (forthcoming in 2008 at http://www.itu.int/md/R07-WRC07-R-0001) (amending ITU Radio Regulations, Article 21) (password protected).

Regulatory Policies

Encouraging research and development of pioneering new technologies is critical, but the resulting technological advances can only take root and flourish if the landscape is properly prepared. Accordingly, the Administration has worked to clear away regulatory obstacles that could thwart the investment that fuels development – and deployment – of new technologies.

Targeted Deregulation to Promote Facilities-Based Investment

The Administration supported the FCC's efforts to modify legacy regulations in order to expand incentives for local telephone companies (incumbent local exchange carriers, or "ILECs") to invest in network upgrades and to stimulate new facilities-based investment by other providers.[28] In a series of orders in its Triennial Review

proceeding, the FCC relieved ILECs of the obligation to make their next generation broadband facilities (*e.g.*, residential fiber-tothe-home (FTTH) and fiber-to-the-curb networks, and FTTH networks serving residential multiunit buildings) available to competitors at discounted rates, terms, and conditions.[29] The Commission determined that these deregulatory measures would make the ILECs more likely to upgrade their existing copper-based facilities "with fiber [that] will permit far greater and more flexible broadband capabilities."[30]

Elimination of Artificial Distinctions among Similar Services

Convergence fueled by broadband services has blurred the lines that differentiated companies in various sectors from one another and served as the historical basis for legacy regulatory models. Cable companies now provide voice and data services over the same lines that carry their video programming services. Telephone companies do the same, as do satellite and wireless providers. With the Administration's backing, the FCC has also moved aggressively to end economic regulation of broadband services. The Commission specifically eliminated disparate treatment for competing providers offering the same types of services, ensuring that broadband services are treated similarly whether they are provided by legacy telephone or cable companies. This has helped to stimulate vigorous competition among these providers.

As an initial step, in 2002, the Commission resolved marketplace uncertainties related to the regulatory status of cable modem services offered by cable television systems. The Commission declared that although these services are interstate in nature, they are "information services" and therefore are neither subject to regulation under Title II of the Communications Act nor to regulation by the states.[31] In 2005, in its order on DSL services, the FCC brought the regulation of DSL services into parity with that of cable modem service by removing legacy regulations from DSL providers that had required them to share their networks with competing Internet service providers (ISPs), tariff their DSL services, and comply with other traditional telephone regulations.[32] In so doing, Chairman Martin observed that the Commission's action constituted:

> an explicit recognition that the telecommunications marketplace that exists today is vastly different from the one governed by regulators over 30 years ago... . [T]he broadband Internet access market today is characterized by multiple platforms that are vigorously competing for customers. Such changed market conditions require, as the Supreme Court in the Brand X decision phrased it, a "fresh analysis." I am pleased that the Commission so quickly undertook this analysis, and, in so doing, removed legacy regulation that applied to only one of the platform providers - the telephone companies.
>
> Broadband deployment is vitally important to our nation as new, advanced services hold the promise of unprecedented business, educational, and healthcare opportunities for all Americans. Perpetuating the application of outdated regulations on only one set of Internet access providers inhibits infrastructure investment, innovation, and competition generally.[33]

Since then, the Commission has established a similar deregulatory framework for broadband Internet access services provided by BPL providers[34] and wireless carriers.[35] These actions reflect the spirit of the Administration's technology-neutral approach. Placing competitors offering similar services on a similar footing creates incentives for investment and helps to ensure that resources are allocated according to competitive merit rather than in response to regulatory constraints.

Voice over Internet Protocol

The Administration has further encouraged aggressive broadband network deployment by supporting policies to ensure that major applications running on the Internet remain free from unnecessary economic regulation. In particular, the

Administration has endorsed FCC policies to date not to impose traditional economic regulation (such as rate regulation, tariffing, and entry and exit regulation) on "voice over Internet protocol" (VoIP) services, and to ensure that states do not impose such regulations. In particular, the Administration strongly supported the Commission's determination that VoIP services that do not utilize the public switched telephone network are unregulated "information services."[36] The Administration also strongly supported the FCC's declaration that VoIP services are interstate services that are not subject to State economic regulation.[37]

Cable Franchise Reform to Stimulate Competition in Video Programming Services

The Administration has also strongly supported efforts to promote competition in the video services marketplace as an important element of broadband policy. The opportunity to deliver competitive advanced programming services creates an incentive for video service providers to deploy broadband infrastructure which, in turn, helps to accelerate broadband penetration. Recognizing this fact, the Administration has, in particular, supported video franchising reform in order to promote competition in the multi-channel video programming market.[38]

Motivated by these same principles, the FCC in 2006 adopted rules to ensure that local cable franchising processes do not unreasonably interfere with rapid broadband deployment or competitors' entry into the cable marketplace.[39] Specifically, the Commission's reforms removed regulatory barriers to new video entrants by, among other things, requiring franchising authorities to act on franchise applications within a reasonable time period and constraining franchise authorities from imposing unreasonable build-out and/or unrelated service requirements as a condition for awarding a franchise.[40] In so doing, the Commission paved the way for more rapid growth of broadband access by removing impediments that had "discourage[d] investment in fiber-based infrastructure necessary for the provision of advanced broadband services."[41]

Opening Federal Land Resources

The Administration has also sought to spur greater broadband deployment by improving the management of rights-of-way on Federal lands. In April 2004, the President signed an executive memorandum that recognized the need of broadband providers for "timely and cost-effective access to rights-of-way so that they can build out their networks."[42] The memorandum has facilitated greater access for service providers to rights-of-way on Federal lands by directing Federal departments and agencies to (1) streamline and standardize applications to reduce the administrative burden and costs on broadband providers, (2) expedite their consideration of rights-of-way applications, (3) ensure that the fees they charge for right-of-way access are reasonable and appropriate, and (4) ensure that they have the proper tools to promote compliance with rights-of-way grants or permits.[43]

FISCAL POLICIES AND TARGETED FUNDING EFFORTS

As his Technology Agenda reflects, the President has long recognized that entrepreneurs and small businesses constitute two of our nation's most important sources for technological innovation. Accordingly, from the earliest days of his Administration, the President has advocated for economic policies to nurture their growth.

In keeping with this view, the Administration endorsed and implemented a number of initiatives specifically designed to stimulate investment in broadband technology. These include both fiscal policies to reduce tax burdens and encourage investment, as well as targeted loan and funding programs to accelerate broadband deployment in rural and remote areas. Coupled with the technology and regulatory policies outlined above, these measures have exerted significant force to advance broadband development and deployment.

Reducing Tax Burdens

The implementation of certain tax provisions has given businesses powerful incentives to invest in broadband technology. Three elements comprise the core of these fiscal initiatives:

Internet Tax Moratorium

From the beginning, preserving the tax-free status of Internet access has been the cornerstone of the President's broadband policy. First enacted in 1998, the Internet tax moratorium prohibits, subject to limited grandfathering, state and local governments from imposing taxes on Internet access and multiple or discriminatory taxes on online transactions. Recognizing that such tax burdens reduce consumer demand for new services, the President has signed extensions of this ban into law three times (2001, 2004, and 2007) and has consistently urged that the moratorium be made permanent.[44] This important legislation provides the marketplace with a predictable environment for investment in new and better technologies while ensuring that the critical communications infrastructure of the Internet remains free of unnecessary taxes.

Accelerated Depreciation

In a rapidly changing area such as the development of new technologies, depreciation policies can also impede investment. New technical advances (and the assets that accompany them) often become obsolete before their value can be recaptured under the tax laws, thereby making investment in new technologies more expensive and uncertain. To combat this problem, the President signed into law, as part of his economic security package (the Jobs and Growth Tax Reconciliation Act of 2003), reforms that allow companies to accelerate depreciation for capital expenditures (including those associated with broadband deployment) and permit the depreciation of as much as 30 percent of the basis in certain property in the first year.[45] These changes have helped stimulate business investment in critical networking infrastructure and broadband equipment.

Tax Credits for Research and Development

Finally, the Administration has promoted tax relief for private research and development. The President signed into law the Tax Relief and Health Care Act of 2006, which extended and enhanced the "Research and Experimentation (R&E) Tax Credit," a long-standing provision of the tax code that provided a 20 percent credit for private research and experimentation expenditures above a certain base amount.[46] Unfortunately, this provision of the code expired at the end of 2007. The President has long advocated making this tax credit permanent and urges Congress to reinstate this provision, which has helped to fuel more innovation across the economy.[47]

FUNDING INITIATIVES AT WORK

Alaska Power & Telephone Takes the High Ground: Bringing Wireless Broadband to the Rural Alaskan Village of Kasaan

KETCHIKAN, Alaska [November 19, 2007] -- Wireless broadband technology that will bring telemedicine, distance learning and other benefits to the people of Kasaan is now a reality with the completion of a successful collaborative effort between Alaska Power & Telephone Company (AP&T), and the Organized Village of Kasaan (OVK).

Completion of the work and delivery of wireless broadband services to Kasaan, an Alaskan Native community located 25 miles west of Ketchikan on Prince of Wales Island, was announced today by AP&T and OVK.

OVK's persistence in pursuing Internet access for their rural community led them to take note of AP&T's success in bridging the digital gap in other rural Alaskan locations. The procurement of $1 million in grant funds under the Federal "*Community Connect Broadband Grant Program*" fueled both hope and a cooperative effort to build the infrastructure necessary for the village to gain high-speed access to the outside world.

Construction on the project began in early 2007. The scope of the project and grant encompassed construction of a mountain-top antenna site above the village on Kasaan Mountain, formation of a community technology center, and delivery of Internet service as well as more than a dozen computers and the other equipment needed to move Kasaan into the digital era. Richard Peterson, President of OVK, states, "AP&T has come alongside us in a manner that has been simply astounding. They have responded to our community in a way few others would."

Certain immediate key benefits to the community will be met by access to broadband technology. These include access to telemedicine, distance learning, connection to other tribal organizations, reduction of travel costs and personal skill-set training in computer and data technologies. It is hoped that by enhancing the local quality of life, business and tribal members will be encouraged to return or stay in Kasaan.

"Coming together with the Village of Kasaan on this project was a natural extension of who we are as an employeeowned Alaskan company, as well as a keystone in AP&T's broader plans for Southeastern Alaska," stated Michael Garrett, Executive Vice President of AP&T and COO of AP&T's Telecom Division.

AP&T provides a broad range of services to more than 30 communities stretching from the Arctic Circle to the southernmost reaches of Southeastern Alaska. These include local and long distance voice, Internet, wireless broadband Wi-Fi, paging, two-way mobile radios, computer networking and delivery of renewable resource based energy services.

Targeted Direct Support for Remote and Rural Areas

To complement the market- based initiatives outlined above, the Administration has also provided targeted seed- funding to support more rapid deployment of broadband in unserved or otherwise under- served rural areas. Since Fiscal Year 2001, USDA has provided nearly $5 billion in loans and grants to support telecommunication expansion and improvements which can improve access to broadband technology.[48] In particular, the USDA supports broadband development in rural America through several programs: the Broadband Loan and Loan Guarantee Program, the "traditional" Telecommunications Infrastructure Loan program, the Community Oriented Connectivity Broadband Grant Program (Community Connect), and the Distance Learning and Telemedicine Program.

Through the Rural Development Broadband Loan and Loan Guarantee Program, USDA provides loans to finance technological solutions tailored to the needs of individual communities. Unlike the Community Connect Grant Program (discussed below), USDA's loans can fund an additional provider in an area. In the seven years of the Program's existence, USDA has approved 70 loans totaling $1.11 billion. The funds have been used to bring a range of technologies including fiber-to-the-premises, DSL, wireless, BPL, and hybrid fiber/coaxial systems to over 1,200 communities in 40 states representing 582,000 household subscribers.[49]

At the time the loans were awarded, over half of these communities either received no broadband service (40 percent) or were served by only one provider (15 percent). To date, eight projects have been completed in Kansas, Louisiana, Michigan, Nebraska, North and South Dakota, Texas, and Washington. In addition, another 38 loans remain in progress (and 15 in initial start-up phase) serving another 987 communities.

In addition, the traditio nal Rural Development Telecommunications Infrastructure Program provides loans to ILECs that are successfully deploying advanced, high-speed networks capable of providing broadband services for their rural customers.[50] Over the past seven years, these borrowers have received over $3.8 billion in loans from USDA to construct and improve facilities designed to offer broadband services.[51] Based on a survey of Rural Development's traditional telephone loan program borrowers conducted in October 2006, approximately 92 percent of these borrowers are providing high-speed Internet service (broadband) to all of the telephone exchanges in their service territories.[52] In addition, they have deployed over 107,000 miles of fiber optic cable.[53]

The Community Connect Grant Program helps ensure that rural consumers receive the same quality and range of telecommunications service as do residents in urban and suburban communities.[54] Under the program, USDA provides financial assistance to furnish broadband service in rural, economically-challenged communities of up to 20,000 inhabitants as yet unserved by a broadband provider. Community Connect grants are awarded on a competitive basis. Grant recipients may use funds to deploy broadband service to "critical community facilities [*e.g.*, rural schools, libraries, education centers, health care providers, law enforcement agencies, and public safety organizations], rural residents, and rural businesses and to construct, acquire, or expand, equip, and operate a community center that provides free access to broadband services to community residents for at least two years."[55]

Over the past five years, USDA has awarded Community Connect Grants totaling $48.3 million for broadband projects.[56] Most re- cently, in October of 2007, USDA awarded $10.3 million to support 19 projects in 13 states.[57] Examples include $834,881 to introduce broadband service in Marcus, Washington, and provide firefighters access to training videos and reports; $603,200 to build a community center and computer lab in Darbyville, Ohio, that will offer high-speed Internet access to universities and medical centers state wide; and $205,416 to build a wireless network in Navajo Mountain, Utah, to provide online public safety notices, educational opportunities, and telemedicine services in this community that is over 100 miles away from the nearest hospital.[58]

Through its Distance Learning and Telemedicine Program, USDA encourages and improves educational and health care services in rural areas through the use of telecommunications, computer networks and related advanced technologies. The program provides funding through 100 percent grants; a combination of grants and loans; or 100 percent loans. Funding amounts start at $50,000. In 2007, USDA awarded a total of $22.3 million to 78 grantees in 31 states.[59] Additionally, USDA awarded $35.9 million in loans under this program to benefit projects in eleven states.

Chapter 3

Progress in Broadband Access

This comprehensive and integrated package of fiscal, technology, and regulatory polices has had a real and positive impact in the broadband marketplace. American consumers are now reaping the rewards of President Bush's vision to "create an environment in which the entrepreneur can flourish [and] in which technologies can reach new frontiers:" a vigorous broadband marketplace in which carriers offering a host of technological platforms compete against one another on price, speed, mobility, content, and other service features.

The lack of a single authoritative data set makes it difficult to establish with certainty whether broadband penetration has become ubiquitous, and this Report acknowledges the benefits of better data gathering tools.[60] Nevertheless, a survey of data from the FCC and other sources reveals a very clear pattern: broadband services have become dramatically more available and affordable across the country. Although additional work remains to be done, broadband penetration is increasing sharply, and consumers – including those in rural and remote areas – have more opportunities than ever to choose the broadband solution (*i.e.*, technology, services, and provider) that best suits their needs and budget.

Vigorous U.S. Platform Growth, Multi-Modal Competition, and Broadband Availability

Perhaps the clearest evidence of the success of the Administration's pro-competitive, technologically-neutral approach lies in the sheer growth in the number of broadband service providers and the broad array of technological alternatives they represent. Over the past few years, the numbers of providers and broadband platforms have proliferated. As Table 1 illustrates, the most current data from the FCC show that the total number of highspeed service providers more than tripled from December 2003 to December 2006 (and increased tenfold between December 2000 and December 2006).

While the data reflect steady increases in the number of high speed cable service providers, and somewhat stronger growth among asymmetrical digital subscriber line (ADSL) providers, it is noteworthy that the most vigorous expansion has occurred among "all other" providers, which include symmetric DSL (SDSL), mobile wireless services, fixed wireless (*e.g.*, Wi-Fi and WiMAX), satellite, fiber optic, and BPL. As Table 1 displays, this category grew by nearly 259 percent, from 246 providers in 2003 to 882 a year ago.

This remarkable growth in the number of broadband service providers has brought with it an even greater increase in the number of broadband lines available to consumers. With this increase in broadband lines, the numbers of subscribers and users have also grown. Chart 1 below reflects that since President Bush took office, the total number of broad-band lines across all modalities has grown by over 1,100 percent, from almost 6.8 million in December 2000 to 82.5 million in December 2006.[61] According to other FCC statistics, over 58 million of these lines serve residential customers.[62]

Competition among the various platforms has also increased sharply. In December 2003, over half (54 percent) of zip codes had three or fewer competitors. (See Table 2.) By December 2006, 91.5 percent of zip codes had three or more competing service providers and more than 50 percent of the nation's zip codes had 6 or more competitors. By contrast, fewer than 3 percent of the nation's zip codes now have one provider or less.

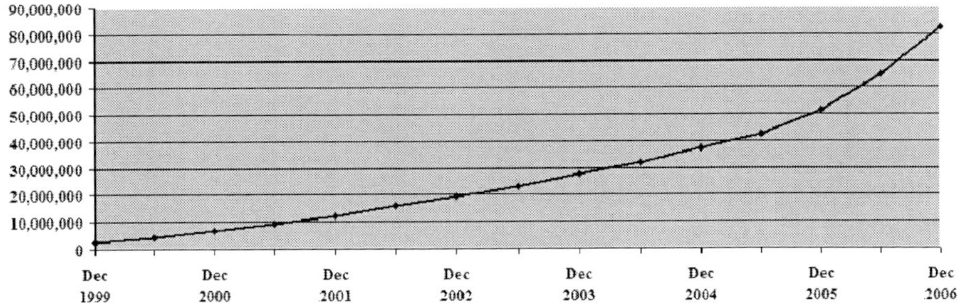

Source: FCC Study, Chart 1

Chart 1. Total High-Speed Lines.

Table 1. Nationwide Number of Providers of High-Speed Lines by Technology (Over 200 kbps in at least one direction)

	ADSL	Cable Modem	All Other [1]	Total
Dec 1999	28	43	65	105
Jun 2000	47	36	75	116
Dec 2000	68	39	87	136
Jun 2001	86	47	98	160
Dec 2001	117	59	122	203
Jun 2002	142	68	138	237
Dec 2002	178	87	169	299
Jun 2003	235	98	217	378
Dec 2003	274	110	246	432
Jun 2004	298	129	281	485
Dec 2004	352	147	312	552
Jun 2005	758	227	779	1,270
Dec 2005	820	242	835	1,347
Jun 2006	832	254	814	1,324
Dec 2006	862	278	882	1,397

For data through December 2004, only those providers with at least 250 lines per state were required to file. Some historical data have been revised.

[1] All other includes SDSL, traditional wireline, fiber, satellite, fixed and mobile wireless, and power line.

Source: High-Speed Services for Internet Access: Status as of December 31, 2006, Federal Communications Commission (October 2007), Table 7 ["*FCC Study*"]

Recent survey data collected by the United States Census Bureau on behalf of NTIA provides further evidence that broadband is experiencing strong growth in the United States. In October 2007, the Census Bureau surveyed approximately 54,000 households, with information on 128,395 persons, as a special Internet use supplement to the Current Population Survey (CPS).[63] Based on its scientific design and size, the survey provides a broad-based and statistically reliable look at Americans' use of the Internet and, more particularly, their use of high-speed technologies to connect to it.

The data reveal that Americans' use of broadband technologies has soared: home broadband usage has risen substantially over the past six years, from 9.1 percent of households in September 2001[64] to 50.8 percent in October 2007.[65] Rural America has also experienced impressive growth. In September 2001, 5.6 percent of rural households had broadband at home;[66] six years later, the rate had reached 38.8 percent, more than a six-fold increase.[67] Naturally, while use of broadband has grown robustly, dial-up use has declined as the significant advantages of an increasingly-available and affordable broadband connection have gained favor.[68] Households in many states have enjoyed rising levels of broadband use: adoption rates today equal or surpass the 40th percentile for at least 41 states and the District of Columbia.[69] New Hampshire and Alaska have usage levels approximating 60 percent of their house- holds.[70]

The foregoing data demonstrate that broadband usage has grown dramatically in the United States. As detailed in the following sections, information from within each of the broadband modal sectors, which shows vibrant growth in deployment and subscriber uptake, reinforces this conclusion. As usage continues to grow, providers will be driven to make further investments to ensure that broadband networks and services are widely available to meet expanding consumer demand.

Table 2. Percentage of Zip Codes with High-Speed Lines in Service

Number of	**1999**	**2000**		**2001**		**2002**		**2003**		**2004**		**2005**		**2006**	
Pro- viders	**Dec**	**Jun**	**Dec**	**Jun**	**Dec**	**Jun**	**Dec**	**Jun**	**Dec**	**Jun**	**Dec**	**Jun**	**Dec**	**Jun**	**Dec**
Zero	40.3	33	26.8	22.2	20.6	16.1	12	9	6.8	5.7	4.6	2	1	0.7	0.4
One	26	25.9	22.7	20.3	19.3	18.4	17.3	16.4	14.9	13.8	12.5	9.3	5.6	3.7	2.4
Two	15.5	17.8	18.4	16.7	15.7	16.2	16.8	16.9	17.1	16.8	16.3	14.1	11.9	8.2	5.7
Three	8.2	9.2	10.9	13.2	13.1	13.3	14.4	14	14.9	14.9	15.1	15	14.8	11.3	8.9
Four	4.3	4.9	6.1	8.2	9.1	9.6	10.3	10.6	11.2	11.6	12.2	12.6	13.5	12.9	11.4
Five	2.7	3.4	4	4.9	6.1	6.9	7.3	7.7	7.8	8.4	8.9	9.7	10.3	12.2	12.5
Six	1.7	2.5	3	3.6	4.2	4.6	5	5.3	5.8	6.1	6.3	6.8	7.8	10.4	11.7
Seven	0.8	1.7	2.3	2.8	3.2	3.2	3.9	4	4.2	4.4	4.6	5.3	5.7	8.7	10
Eight	0.3	0.8	2	2.2	2.5	2.8	2.7	3.1	3.3	3.6	3.6	4	4.6	7.1	8.3
Nine	0.2	0.4	1.6	1.9	2	2.4	2.2	2.5	2.6	2.8	3.1	3.8	4	5.8	6.7
Ten or More	0	0.4	2.4	3.9	4	6.4	8	10.5	11.4	11.8	12.8	17.5	20.7	19.1	22

For data through December 2004, only those providers with at least 250 lines per state were required to file. Figures may not add up to 100% due to rounding.

Source: *FCC Study*, Table 15.

DIGITAL SUBSCRIBER LINES (DSL)

Wireline communications service providers, such as ILECs and competitive local exchange carriers (CLECs), primarily offer broadband services using DSL, a technology that allows them to derive a separate high capacity transmission channel from the copper loop used to provide conventional voice service. There are two basic varieties of DSL service – "asymmetric," for which the transmission speed in one direction (typically the "downstream" path from the network to the customer) is faster than the other, and "symmetric," for which the transmission speed is equivalent in both directions.

Industry data demonstrate significant growth in the availability and affordability of DSL services. According to the FCC, as of year end 2006, broadband DSL was available to 79 percent of the households in areas where ILECs offer local telephone service.[71] More than 82 percent of the homes served by the largest ILECs – Verizon, AT&T, and Qwest – can obtain DSL service.[72]

Importantly, evidence suggests that this progress is not restricted to the more populated areas of the country. An April 2007 survey by the Organization for the Promotion and Advancement of Small Telecommunications Companies (OPASTCO) revealed that ILECs offer broadband services to more than 90 percent of their rural customer base, chiefly via DSL.[73] Similarly, based on December 2006 data, the National Exchange Carrier Association (NECA) found that 511 of its member companies (representing 462,000 DSL lines located in 47 states and Guam) averaged 91 percent DSL availability.[74] In the study, 352 of the companies offered DSL to between 91 and 100 percent of their customers, while another 80 provided such coverage to between 81 and 90 percent of their customer base. In all, 84 percent of the companies made DSL available to more than 80 percent of their customers.[75] ILECs responding to a 2006 survey by the National Telecommunications Cooperative Association (NTCA) indicated that all offered broadband service to some part of their rural customer base, up from 58 percent of the respondents in 2000. Again, DSL was the primary broadband service offering.[76]

As DSL service has been extended to more and more households, service speeds have accelerated and prices have dropped. According to Verizon, in the past three years, the major DSL operators have increased their top downstream speeds from under 1 Mbps to 2- 3 Mbps.[77] AT&T notes that wireline providers currently offer their DSL customers a variety of transmission speeds, typically ranging from 768 Kbps up to 6 Mbps.[78] Ninety percent of the responding rural ILECs in the OPASTCO survey could deliver data speeds of at least 1 Mbps in one direction.[79]

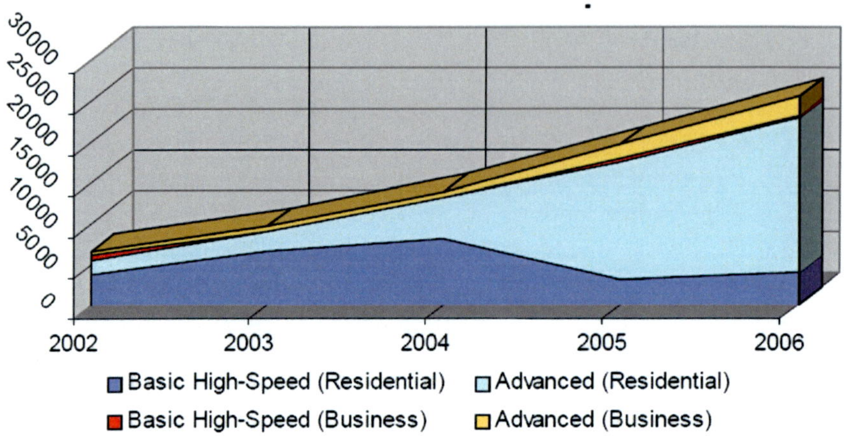

Source: NTIA (data from FCC Study, Tables 1-4).

Note: Total High-Speed Lines consist of "Basic" High-Speed plus "Advanced" Service Lines.

Chart 2. ADSL Growth 2002-2006 No. of High-Speed Lines (Thousands).

The increasing speed of DSL services has not been accompanied by increased costs for consumers. By one estimate, the average monthly revenue per user of DSL service decreased from $40 in 2002 to $31 in 2006.[80] Between 2004 and 2007, AT&T cut its monthly rate for 1.5 Mbps DSL service nearly 60 percent, from $49.95 to $19.99.[81] According to the FTC, between May 2005 and April 2006, the price of AT&T's 3.0 Mbps DSL offering dropped from $29.95 to $17.99.[82]

With availability and speeds up and prices down, DSL subscribership has burgeoned. According to the FCC's most recent statistics, there were about 22.8 million residential "high-speed" ADSL lines in service as of year end 2006, up over 300 percent from December 2002.[83] (See Chart 2.)[84]

Adding commercial customers would increase the number to some 25.4 million.[85] The FCC data depict even larger subscriber gains for "advanced" ADSL services; there were nearly 19 million residential advanced lines in service at year end 2006, up more than 933 percent from December 2002.[86] Including commercial users would boost the subscriber count to more than 21.1 million lines.[87]

Evidence also suggests that this growth in DSL deployment has been fueled by FCC decisions reducing competitors' rights to gain access to ILECs' next generation networks and relaxing regulation of ILECs' broadband services.[88] Professor Thomas Hazlett has shown that DSL subscribership grew more rapidly after the FCC's 2003 Triennial Review Order than prior year growth trends would have suggested.[89] (See Chart 3) Similarly, Jeffrey Eisenach of Criterion Economics has argued that the FCC's deregulatory decisions since 2003 have stimulated investment in communications equipment, helping the industry recover from the technology sector "meltdown" of 2001-2002.[90]

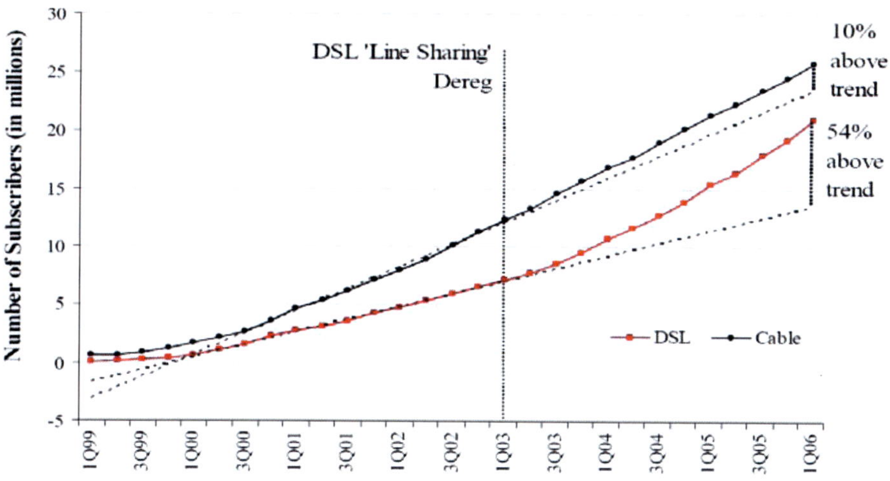

Source: Legg Mason.

Source: T. Hazlett, Legg Mason, http://mason.gmu.edu/~thazlett/Conferences/LessonsfromtheTelecomWarsPresentations.html

Chart 3. DSL and Cable Modem Subs, 1Q1999 - 1Q2006.

High-Speed Cable

Data from the cable industry also reflect significant growth in deployment and consumer uptake of cable-delivered high-speed data services. The FCC's statistics, set forth in Table 1, above, indicate that the total number of cable modem service providers grew by over 150 percent in the three years from December 2003 to December 2006 (rising from 110 to 278).

Cable industry research conducted by Kagan Research (now SNL Kagan) reveals that, during that same interval, cable operators increased the number of housing units they pass with broadband service by 15.9 million. (Chart 4) Based on these figures, NCTA estimates that high-speed cable service currently reaches 92 percent of all U.S. households as of the end of 2007.[91]

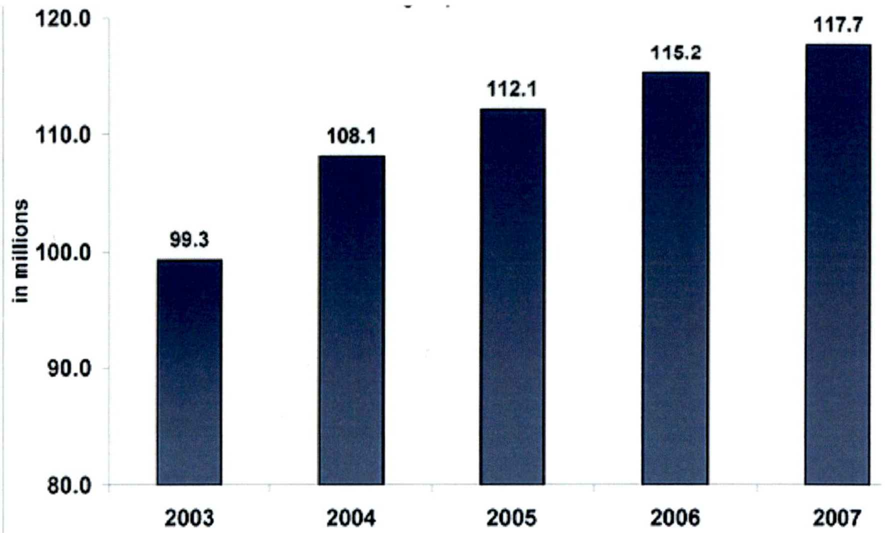

Note: Total housing units passed figures have been adjusted to avoid double counting of estimated housing units passed by both incumbent cable operators and overbuilders.
Source: *NCTA* (SNL Kagan) (used with permission) http://www.ncta.com/Statistic/Statistic/CableBroadbandAvailability.aspx

Chart 4. Housing Units Passed by Cable High-Speed Internet Service.

This same research also reflects significant increases in uptake of broadband services. At the end of 2006, the cable platform was delivering high-speed data services to over 31 million customers – an increase of over 125 percent from the 13.8 million customers served in the second quarter of 2003. (Chart 5) Residential customers comprised 30.2 million of these high-speed data customers. According to SNL Kagan, by mid-year 2007, the total number of cable high-speed data customers had grown to 33.5 million.

Accompanying this growth in subscribers has been an increase in the performance cable operators are delivering to their customers. The industry's trade association reports that most cable broadband systems now offer transmission speeds that exceed 5 Mbps, with some delivering as much as 50 Mbps.[92] With the growth of competition across technical platforms, evidence suggests that cable operators are developing next generation wideband systems capable of delivering transmission speeds of over 100 Mbps in order to respond to the competitive challenge presented by other alternative broadband providers.[93] These actions, in turn, exert competitive forces on others.

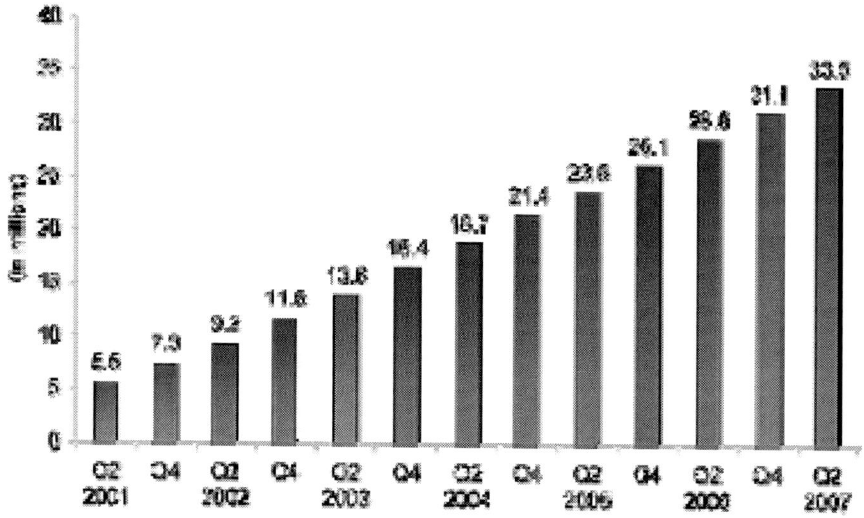

Note: Figures include business cable modem customers; the most recent estimate of residential-only HSI customers is 30.2 million at 4Q06.

Source: *NCTA* (SNL Kagan) (used with permission) http://www.ncta.com/Statistic/Statistic/ResidentialCableHighSpeedDataSubscribers.aspx

Chart 5. Cable HIS Customers 2001-2007.

For example, exemplifying the importance of facilities-based competition emphasized in the Commission's Triennial Review Order, cable operators have leveraged their existing broadband capabilities to seize a significant share of the residential landline market.[94] The cable industry reports that it had 9.5 million VoIP subscribers in 2006, a 61 percent increase over a year earlier.[95] According to the Telecommunications Industry Association (TIA), cable VoIP services held almost 10 percent of that market in 2006, and the share is expected to exceed 33 percent by 2010.[96] Noteworthy in its own right, this development holds even greater significance for broadband deployment and accessibility. To compete with cable's bundled offerings, landline telephone operators have accelerated their deployment of fiber capacity in order to add high definition and multi-channel video and other bandwidth-dependent entertainment applications to their own service offerings.[97] This trend is detailed more fully below in the section on fiber.

MOBILE WIRELESS

The wireless industry is the fastest growing sector of America's broadband economy. According to industry data, from 2000 to 2007 the total number of subscribers increased from 97 million to 243 million, and the revenues in the industry grew from $45 billion to $132 billion. (*See* Table 3, Chart 6)

Fueled in large part by demand for non-voice applications (*e.g.*, video services, multimedia and text messaging, wireless games, and music), mobile broadband services have contributed significantly to the growth of the mobile wireless sector.[98] The FCC's most recent data reveal that the number of broadband lines provided by mobile wireless operators increased from less than 380,000 in June of 2005 to almost 22 million at the end of 2006 – a growth rate that dwarfs that of other broadband platforms.[99] (Chart 7) Moreover, a significant portion of these lines (4.1 million, or almost 19 percent) provide "advanced" services capable of delivering over 200 Kbps in both directions.[100]

Table 3. Annualized wireless industry survey results - June 2000 to June 2007 Reflecting Domestic U.S. Commercially-Operated Cellular, ESMR and PCS Providers

Date	Estimated Total Subscribers	Twelve-Month Total Service Revenue (in $000s)	12-Month Roamer Revenues (in $000s)	Cell Sites	Direct Service Provider Employees	Average Local Monthly Bill (June Survey Periods)	Average Local Call Length (June Survey Periods)
2000	97,035,925	$45,295,550	$4,134,626	95,733	159,645	$45.15	2.48
2001	118,397,734	$58,726,376	$3,698,683	114,059	186,317	$45.56	2.62
2002	134,561,370	$71,117,599	$3,872,035	131,350	186,956	$47.42	2.6
2003	148,065,824	$81,185,272	$3,874,488	147,719	187,169	$49.46	2.63
2004	169,467,393	$95,515,593	$3,956,823	174,368	212,186	$49.49	3.06
2005	194479364*	$108,534,727	$4,136,492	178,025	225,162	$49.52	3.04
2006	219652457*	$118,299,682	$3,558,052	197,576	238,236	$49.30	2.94
2007	243,428,202	$132,893,824	$3,611,049	210,036	257,401	$49.94	3.13

* Restated.

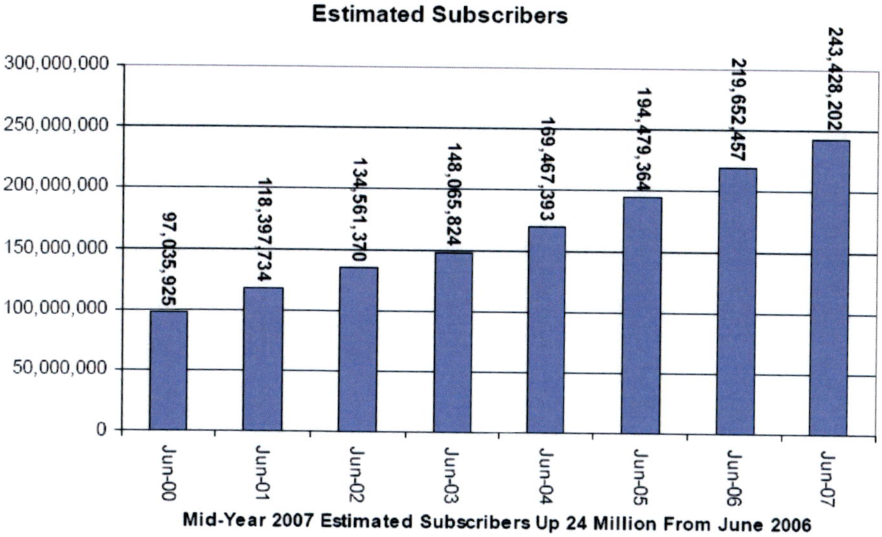

Source: Semi-Annual Wireless Industry Survey, 2007, CTIA (used with permission).

Chart 6. Estimated Subscribers.

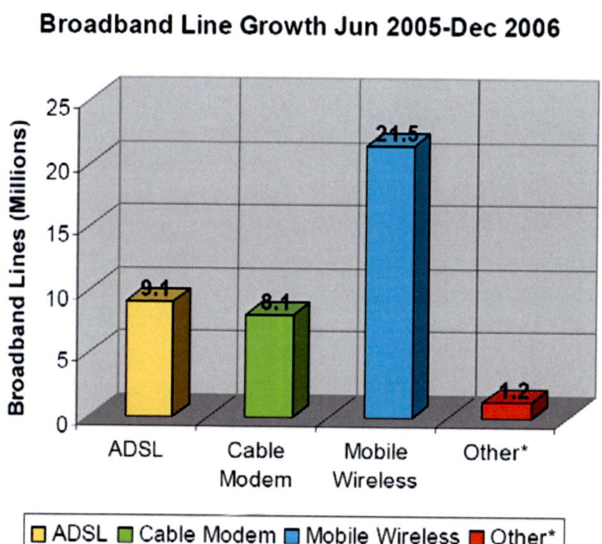

* Includes Fiber, Satellite, Fixed Wireless, and Power Line.
Source: NTIA (data from FCC Study, Table 1).

Chart 7. Broadband Line Growth Jun 2005-Dec 2006.

On an inter-modal basis, mobile wireless competes primarily on the basis of its convenience and portability. However, to a somewhat greater extent than other broadband sectors, the mobile wireless market also exhibits a significant degree of intra-modal competition. As four major national service providers – AT&T, Sprint Nextel, T-Mobile, and Verizon Wireless – and a score of regional providers compete aggressively for customers, all of the carriers face pressure to introduce new and attractive features and services on a rapid basis. In this context, broadband applications (and the faster transmission speeds they require) constitute an important battlefront. Spurred by competitive pressures, each of the major carriers has taken steps to develop and deploy third generation (3G) architectures to deliver these advanced services and features.

The FCC's analysis confirms this trend. In its most recent analysis of competition in the commercial mobile radio services (CMRS) marketplace, the Commission observed that "the record indicates that competitive pressure continues to drive carriers to introduce innovative pricing plans and service offerings, and to match the pricing and service innovations introduced by rival carriers."[101] Moreover, with respect to mobile broadband services, the FCC 11th Annual CMRS Competition Report found that:

> [i]n addition, the deployment of next- generation networks based on competing technological standards continues to be an important dimension of non-price rivalry in the U.S. mobile telecommunications market. In December 2005, Cingular Wireless [now AT&T] commercially launched UMTS [universal mobile telecommunication system] (or WCDMA [wideband code-division multiple access]) with HSDPA [High-Speed Downlink Packet Access] in 16 U.S. cities to compete with the EV-DO [Evolution Data Optimized]-based wireless broadband services previously launched by Verizon Wireless, Sprint Nextel, and some regional CDMA carriers such as Alltel. Because the speeds on EV-DO and WCDMA/HSDPA networks are much faster than the speeds on European WCDMA networks, it has been argued that the deployment of these next-generation technologies by U.S. wireless carriers has given the United States an edge over Europe in wireless data networks for the first time in years.[102]

The FCC's study, released in September 2006, examined data from 2005 and early 2006. More recent information from the companies themselves reflects the significant progress in these deployment activities that has transpired in the last year.

For example, AT&T reports that its HSDPAbased BroadbandConnect network is now available in over 200 major metropolitan areas in the United States and provides average download speeds of 600 Kbps to 1.4 Mbps.[103]

Verizon Wireless recently announced that it has upgraded its entire EV-DO network to Revision A (Rev. A) technology, making average download speeds of 600 Kbps to 1.4 Mbps and average upload speeds of 500-800 Kbps available to over 210 million people across the United States.[104] Sprint Nextel similarly states that its EV-DO Rev. A service offers download speeds of 600 Kbps to 1.4 Mbps, and upload speeds of 3 50-500 Kbps.[105]

Its network presently reaches 212 million people, and the company expects to reach up to 280 million people (92 percent of the nation's population) by the end of 2008.[106]

T-Mobile, the fourth major national wireless service provider, is preparing to deploy its own HSDPA-based 3G network infrastructure in the first half of 2008.[107] T-Mobile's network will utilize the 1.7 GHz/2.1 GHz spectrum the company acquired in the AWS auction (Auction No. 66) in 2006 – which effectively doubled its available spectrum assets in the top 100 markets – and will deliver download speeds of 0.8 Mbps to 1.1 Mbps.[108] In addition, T-Mobile – among the nation's largest operators of fixed wireless public access points (hot spots) – in 2007 launched the first nationwide commercial unlicensed mobile access (UMA) based service in the United States. The service, T-Mobile Hot-Spot@Home, is discussed more fully in the section on the fixed wireless sector below. [109]

The diversity of competing technological standards (*i.e.*, UMTS, EV-DO, HSDPA) that has emerged in the wireless broadband market can be traced directly to technology-neutral regulatory policies intended to foster innovation. Likewise, T-Mobile's forthcoming 3G deployment in the AWS band is an immediate consequence of Federal efforts to make additional spectrum available for commercial uses.

FIXED WIRELESS

Fixed wireless technologies – *i.e.*, wireless systems or devices that are deployed in fixed locations as distinguished from mobile devices such as cell phones or personal digital assistants – have also emerged as an important complement to the mobility afforded by CMRS,[110] and as a potential "last-mile" solution to deliver broadband to currently unserved areas[111] According to FCC figures, the number of fixed wireless broadband lines in the United States grew 132 percent between June 2005 and December 2006 – from 208,695 to 484,073.[112] TIA estimated the total number of fixed wireless subscribers in 2006 to be 800,000.[113]

As outlined below, fixed wireless technologies have contributed significantly to the rapid growth of broadband accessibility and usage in the United States, and recent technological advances position the sector for further growth.[114] The two principal technologies in this market segment are Wi-Fi and WiMAX.[115]

Wi-Fi

In its 2006 CMRS Competition Report, the FCC noted that

> Wireless Local Area Networks (WLANs) and Wireless Metropolitan Area Networks (WMANs) are playing an increasingly important role as a competitor and supplement to the services offered by the CMRS industry. WLANs are already widely deployed and enable consumers to obtain high-speed wireless Internet connections within a range of 150 to 250 feet from a wireless access point (AP). The most prevalent WLAN technology is equipment manufactured in accordance with the IEEE 802.11 family of standards, commonly known as "Wi-Fi," short for wireless fidelity. Basic WLAN data transfer rates range from speeds of up to 11 Mbps for 802.11b and up to 54 Mbps for 802.11a and 802.11g. New "SpeedBoost" or "Super G" routers, marketed as "pre-802.11n," employ MIMO (Multiple Input Multiple Output) technology, making them capable of providing speeds from 108 to 240 Mbps.[116]

The increasing penetration of broadband DSL and cable, discussed above, has stimulated a commensurate climb in Wi-Fi adoption as people deploy home WLANs to share broadband connections among several computers.[117] Moreover, the proliferation of Wi-Fi access points, or "hot spots," in public venues such as book stores and coffee shops has increasingly enabled consumers to cut the tether to a home or office connection altogether.

And the number of hot spots has grown – considerably. According to JiWire.com, the United States currently ranks first in the world in the number of public/commercial Wi-Fi hot spots, with an estimated 66,058 – more than double the next closest country. (Table 4) An earlier estimate from TIA placed the figure at 53,000 in 2006.[118] TIA noted that this reflected an almost 36 percent increase from 39,000 in 2005, and TIA projects that the number will grow further to 83,000 in 2010.[119]

Table 4. Top 10 Countries for availability of Wi-Fi Service

Top 10 Countries for availability of Wi-Fi Service	
United States	66,058
United Kingdom	31,055
France	22,959
Germany	21,347
South Korea	15,635
Japan	8,348
Spain	5,792
Italy	5,133
Taiwan	4,391
Switzerland	3,108

Source: JiWire WiFi HotStatsTM (as of December 21, 2007) http://www.jiwire.com/search-hotspot-locations.htm

Competition in the wireless hot spot marketplace is brisk. According to TIA, AT&T presently operates approximately 11,000 hot spots in the United States;[120] T-Mobile operates 8,600, with over 700 more accessible under the company's roaming agreements;[121] and a range of providers including Boingo, Wayport, Fiberlink, and ipass, also provides access through wireless hot spots.

With this rise in penetration and usage, Wi-Fi has also become a fertile ground for technical innovation and entrepreneurship. With technology neutral policies as a catalyst, engineers and service providers seeking to strengthen their competitive position have devised system architectures that employ Wi-Fi technology in new and creative ways to exact maximum value from available spectrum resources and deliver it to consumers. Two examples illustrate this phenomenon well: the emergence of UMA technology and municipal Wi-Fi deployments.

UMA technology, as its name implies, facilitates access to commercial cellular networks (in particular, those using global systems for mobile communication (GSM) and general packet radio service (GPRS)) using unlicensed spectrum technologies such as Bluetooth and Wi-Fi.[122] Using dual-mode mobile handsets, UMA allows subscribers to roam and hand- off calls between public and private unlicensed wireless networks and conventional cellular networks.[123]

As noted above, T-Mobile has been an early advocate for UMA technology in the United States. Its "Hotspot @Home" service, launched in 2007, supports both voice and multi-media applications[124] and enables users to connect calls directly through the Internet from a home wireless network or public hot spot using voice over Internet Protocol (VoIP).[125] Outside of a Wi-Fi coverage area, handsets can function as conventional cellular phones using the carrier's mobile wireless network.[126] In the highly competitive wireless sector, technology firms Qualcomm and Nortel recently announced that they are collaborating on a competing "fixed-wireless convergence" technology, IP Multimedia Subsystem (IMS) Voice Call Continuity.[127] These innovative approaches to bridging the gap between traditional CMRS and fixed wireless services have thrived under the current regulatory scheme.

Municipal Wi-Fi efforts also evince the innovative use of Wi-Fi technology to expand broadband access for the American people. In its study of broadband competition released in 2007, the Federal Trade Commission (FTC) took notice of the increasing interest of many municipalities in Wi-Fi fixed wireless technology as a less expensive means to deliver broadband Internet access to their residents.[128] Observing that "[l]ow-cost municipal broadband services reduce the digital divide separating more affluent consumers from less affluent ones," TIA estimated that over 250 municipal Wi-Fi networks are being operated or developed throughout the United States.[129] A more recent study by ABI Research places this number at over 400.[130] The projects employ a variety of models including public-private partnerships, contracted services, and municipally-owned facilities.[131]

While some early municipal Wi-Fi projects, specially in larger cities, have encountered difficulties,[132] systems are also taking root in smaller communities, where smaller geographic scale can prove to be an advantage.[133] For example, in Alamance County, North Carolina, the town of Graham has budgeted $80,000 in 2007-08 to develop a system that would support local police and provide free Internet access.[134] Fueled by a $190,000 grant from USDA, Pinal County, Arizona, also recently announced that it will deploy a Wi-Fi mesh network to provide broadband service to the rural outskirts of the county by 2009,[135] and in Glendora, Mississippi, a small northwestern Mississippi Delta village of 285 people, a $325,405 USDA grant is being used to develop a fixed wireless technology project that will bring broadband connectivity to the community. The project will connect the public library, health clinic, police department, and volunteer fire department.[136] Such small communities' municipal Wi-Fi efforts have often enjoyed greater success than those in large cities,[137] and many analysts believe that the new business model reflected in some of these projects, which provides broadband access for residents by leveraging a Wi-Fi system built primarily to support municipal government services, represents the best future for municipal Wi-Fi.[138]

WiMAX

As the inventive use of Wi-Fi has contributed to expanding broadband penetration and usage in the United States, WiMAX offers yet another emerging technological tool with the potential to deliver even greater gains in broadband accessibility in the future. WiMAX can deliver fixed wireless broadband access across much wider geographical areas than Wi-Fi; covering distances as great as five miles without line of sight and up to 30 miles under ideal conditions.[139]

With potential data speeds up to 70 Mbps, WiMAX has been identified as a possible "last-mile" solution to deliver broadband into rural and remote areas where high-speed cable service and DSL are not available.[140] Verizon Communications has been testing fixed WiMAX for this purpose.[141] AT&T has also been evaluating WiMAX technology with test deployments in Alaska, Georgia, Nevada, New Jersey, and Texas.[142] However, WiMAX is also being developed for mobile applications and, in October 2007, the International Telecommunication Union approved WiMAX for inclusion in the International Mobile Telecommunications 2000 (IMT-2000) family of standards for next generation mobile wireless communications.[143]

In the United States, Clearwire Corporation of Kirkland, Washington, has been among the earliest and strongest proponents for WiMAX technology, for both fixed and mobile services. Clearwire commenced service in its first market in August 2004.[144] Clearwire now serves 53 urban and rural markets in 16 states,[145] and served approximately 258,000 subscribers as of the end of the first quarter of 2007 – a 161 percent expansion over a year earlier.[146] Its present service uses pre-WiMAX, or "WiMAX-ready," proprietary non-line-of-sight orthogonal frequency division multiplex (OFDM) technologies operating in the 2495-2690 MHz band to deliver wireless broadband access at downstream transmission speeds up to 1.5 Mbps.[147] Clearwire reports, however, that it intends to deploy full WiMAX networks (*i.e.*, compliant with IEEE standard 802.16e-2005) for fixed, portable, and mobile applications when mobile WiMAX equipment becomes commercially available, and the company has established strategic partnerships with Intel and Motorola to develop that equipment.[148]

Indeed, reports indicate that Intel has built its wireless strategy around the mobile WiMAX standard.[149] This past autumn, the company announced that it will launch its next generation Centrino mobile platform, called Montevina, in the middle of 2008.[150] Montevina will include an integrated Wi-Fi/WiMAX module, and Intel has already secured commitments from several laptop makers, including Acer, Lenovo, Panasonic, and Toshiba, to use Montevina.[151] This penetration is expected to increase the potential market for WiMAX service.[152] Reflecting similar confidence in the growth potential for WiMAX, Cisco Systems, Inc., recently announced plans to spend $330 million to acquire Navini Networks, a developer of technologies to enhance the range and performance of WiMAX-based services.[153]

Working with Intel and Motorola, Sprint Nextel has also built its next-generation strategy on mobile WiMAX technology.[154] In August 2006, it announced its intention to employ mobile WiMAX as the foundation for its fourth generation (4G) wireless broadband network, which will use the company's considerable spectrum resources in the 2.5 GHz band.[155] Sprint Nextel had indicated that it would invest up to $5 billion to deploy the new network and planned to reach a potential 100 million customers by the end 2008.[156] Although it is reexamining that investment budget and its original deployment schedule, the company has said it remains committed to its WiMAX 4G strategy and plans to move forward with a "soft launch" of WiMAX service in the Baltimore-Washington and Chicago areas.[157]

SATELLITE

Satellite technology also represents an important conduit for the delivery of broadband access to the nation. Indeed, the FCC's most current data show satellite to be the most ubiquitous of the nation's broadband technologies, delivering service to at least one customer in 91 percent of U.S. zip codes at the end of 2006.[158]

At present, satellite broadband is primarily provided utilizing one or two geostationary satellites (satellites with high orbits that remain in the same virtual point in the sky) to deliver service to subscribers using small receiving "dishes" at fixed locations. Fixed satellite service (FSS) providers such as Wild- Blue Communications, Inc., Hughes, and Gilat (operator of the StarBand service) have been able to blanket virtually the entire United States with near DSL speeds.[159] (Table 5). Inmarsat, through its Broadband Global Area Network (BGAN) service, uses its constellation of geostationary satellites to deliver integrated voice and satellite broadband access as a mobile service to customers using portable terrestrial equipment, though at comparatively lower speeds. [160] (Table 5)

Table 5.

Satellite Broadband Company / Service Offering	'Up to' Upload Speed	'Up to' Download Speed	Monthly Service Price	Satellite CPE Cost	Installation Cost
Hughes (www.hughes.com)					
HughesNet Home Service	128 Kbps	700 Kbps	$59.99	$299.99	Included
HughesNet Professional	200 Kbps	1.0 Mbps	$69.99	$299.99	Included
HughesNet Professional Plus	200 Kbps	1.5 Mbps	$79.99	$299.99	Included
HughesNet Business for Small Office	300 Kbps	1.5 Mbps	$99.99	$599.98	Included
HughesNet Business Internet	500 Kbps	2 Mbps	$179.99	$599.98	Included
StarBand (www.starband.com)					
Residential	128 Kbps	1.0 Mbps	$69.99	$299.99	Not Included
Small Office	256 Kbps	1.5 Mbps	$99.99	$299.99	Not Included
WildBlue (www.wildblue.com)					
WildBlue for Home Value Pak	128 Kbps	512 Kbps	$49.95	$249.00	Included*
WildBlue for Home Select Pak	200 Kbps	1.0 Mbps	$69.95	$249.00	Included*
WildBlue for Home Pro Pak	256 Kbps	1.5 Mbps	$79.95	$249.00	Included*
WildBlue for Office Select Pak	200 Kbps	1.0 Mbps	$69.95	$249.00	Included*
WildBlue for Office Pro Pak	256 Kbps	1.5 Mbps	$79.95	$249.00	Included*

* In some cases , WildBlue charges a $79.95 installation fee.

Inmarsat (www.inmarsat.com)			
Inmarsat BGAN	492 Kbps	492 Kbps	Pricing depends on individual distributor's offering and what value-added services are included

Source: Satellite Industry Association (compiled from industry websites).

Like its earthbound competitors, satellite broadband has seen tremendous growth over the past few years. From fewer than 50,000 subscribers in 2004, satellite broadband providers were serving an estimated 700,000 subscribers by 2006.[161] Hughes Network Systems reported that it had over 350,000 residential subscribers as of October 2007,[162] and was adding more than 11,000 new customers each month.[163] The ubiquity of its coverage, makes satellite well-suited to deliver service to rural, remote, and otherwise under- served areas of the country normally not covered by traditional sources of broadband Internet services such as cable, DSL, fiber, or wireless.[164]

Next-generation satellite broadband offerings by TerreStar,[165] Inmarsat,[166] and Mobile Satellite Ventures (MSV)[167] are also in development. Inmarsat and TerreStar both plan to launch new satellites in mid-2008, which will greatly enhance availability of broadband services, number of simultaneous users, and total bandwidth. By incorporating more powerful transmitters and an ancillary terrestrial component (ATC), next generation providers will be able to offer services that more closely resemble those of terrestrial fixed and mobile wireless services.[168]

Fiber Optic Cable and Broadband over Power Lines

Fiber optic cable and electric power transmission and distribution lines represent technological alternatives to other traditional wireline-based broadband platforms. In its May 2006 report on broadband deployment, the Government Accountability Office (GAO) characterized "deep fiber deployment" and BPL as potentially important broadband delivery platforms "in the coming years."[169] However, the most recent data from the FCC and other sources reflect more rapid emergence of these technologies.

According to the Commission's most recent study, the total number of high speed lines delivered over fiber or power line connections grew 789 percent from December 2003 to December 2006, rising from 116,390 to just over a million.[170] Data from June 2005 (the first point at which fiber and power line statistics were disaggregated) forward show that optical fiber deployments to subscribers' premises (i.e., FTTH) comprise the largest portion of this expansion.[171]

Table 6. Annual Fiber Deployment in the United States (Thousands of Miles)

Year	*Telephone Companies*	*Cable System Operators*	*Utilities/ Municipalities*	*Private Campuses/ Buildings*	*Total*
2000	14,435	3,505	839	777	19,556
2001	10,009	4,163	932	652	15,756
2002	3,224	1,926	466	373	5,989
2003	2,464	1,367	516	466	4,813
2004	5,567	1,212	606	435	7,820
2005	8,816	994	746	466	11,022
2006	9,737	994	808	482	12,021
2007	10,666	1,056	870	497	13,089
2008	11,741	1,118	932	528	14,319
2009	12,434	1,181	994	559	15,168
2010	12,972	1,243	1,056	590	15,861

Source: TIA Market Study, Table II-4.1 (used with permission).

Fiber Optic Cable

Data compiled by industry sources also document the steady expansion in fiber growth evident in the Commission's findings not merely for FTTH deployments, but also elsewhere in the providers' networks. TIA has charted patterns in fiber deployment dating back more than a decade.[172] Its data show that, following a steep decline from 2000 to 2003 to absorb overcapacity from earlier build-outs, annual deployment of fiber in the United States grew sharply over the last four years, from 4.8 million miles fielded in 2003 to 13.1 million miles in 2007.[173] Moreover, TIA expects deployment to continue to rise, with operators laying an estimated 58.4 million miles from 2007 to 2010.[174] (*See* Chart 8)

TIA's data further reveal that telephone companies have led the way in this resurgence, nearly quadrupling their deployment of fiber from 2003 to 2006.[175] (*See* Table 6)

As noted in the cable discussion above, telephone companies' aggressive deployment of fiber has been motivated by the competitive need for a platform that will enable them to deliver higher-speed data services and digital, high-definition television in order to compete with cable operators' bundled packages.[176] For this reason, TIA expects fiber deployment and accessibility to continue to grow as these companies build out FTTH (also called fiber-to-the-premises (FTTP)) and fiber-to-thenode (FTTN) networks.[177]

As noted above, FTTH systems consist of fiber optic transmission facilities running from the telephone company's central office to the customer's premises.[178] FTTH gives the service provider significantly more capacity than traditional copper infrastructure and supports significantly higher transmission speeds (between 100-500 Mbps).[179] Like the other broadband platforms discussed above, FTTH has experienced remarkable expansion. A recent study prepared by RVA Market Research for the Fiber-to-the-Home Council (FTTH Council) and TIA shows that FTTH networks now pass 9.55 million homes in North America (TIA indicates that virtually all of these homes are in the United States), a more than 50 percent increase from September 2006.[180] Subscribership – at 2.14 million homes – also grew by nearly 112 percent over that period.[181] Perhaps most significant, however, is the fact that the "take rate" (the number of households that actually subscribe after being offered the service) has increased to 26.8 percent from 22.3 percent only six months earlier. [182]

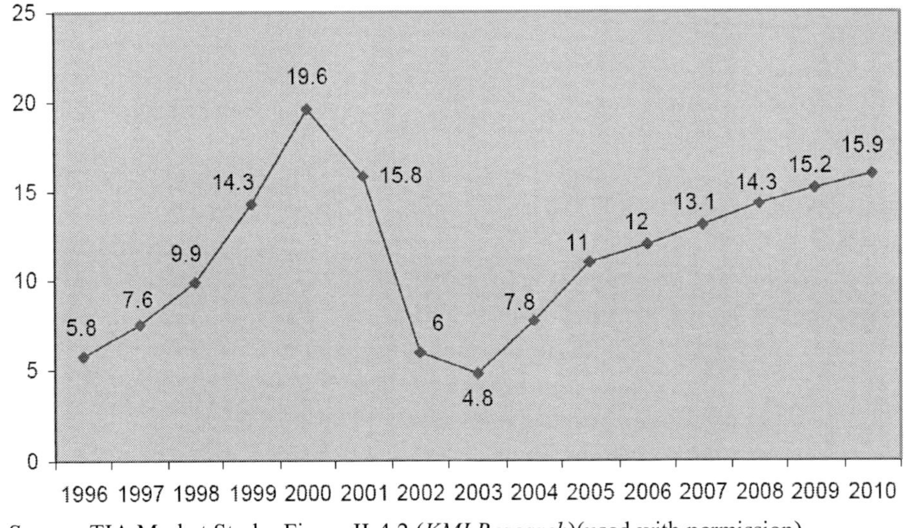

Source: TIA Market Study, Figure II-4.2 (*KMI Research*)(used with permission)

Chart 8. fiber Miles Deployed in the United States (Millions).

At present, there are approximately 370 providers delivering FTTH services, including public utilities, municipalities, small and medium size telephone companies, and cable operators.[183] However, Verizon Communications remains the primary operator.[184] Verizon Communications' fiber-to-the-premises network (FiOS) delivers asymmetric connections of either up to 50 Mbps (upstream)/20 Mbps (downstream) or up to 30 Mbps/15 Mbps depending on the state where the service is sold.[185] As of November 2007, Verizon began offering high speed symmetric FiOS Internet services to consumers in 16 states with connections of 15 Mbps or 20 Mbps depending on the state where the service is sold.[186] At the end of September 2007, Verizon had passed about 8.5 million homes and businesses in 16 states with its FTTP network, and the company plans to pass 3 million premises annually with its FTTP network through 2010.[187]

In contrast to Verizon's FiOS network, which runs fiber end-to-end (from Verizon's central office to the customer's premises), AT&T is investing heavily in developing a network that supplements some FTTP capacity with FTTN infrastructure.[188] FTTN employs fiber from the central office to neighborhood "nodes" 3,000 to 5,000 feet from the customer's home or business, but uses existing copper lines for the last mile between the node and the customer's premises.[189]

AT&T plans to deploy approximately 40,000 miles of new fiber to enhance its existing network.[190] The platform will support AT&T's "U-verse" package of services, which includes an IP-based digital video service (IPTV); U-verse enabled high-speed Internet access offering speeds ranging from 1.5 to 6.0 Mbps downstream and 1.0 Mbps upstream; and, ultimately, VoIP service as well.[191] As of the end of the third quarter of 2007, AT&T served approximately 126,000 U-verse TV and Internet subscribers, and passed about 5.5 million homes with its service.[192] However, the company was adding approximately 10,000 subscribers per week at that time and expects to expand its service area to cover 17 million homes by the end of 2008.[193]

Build-out of the FiOS and U-verse platforms is expected to dominate fiber deployment over the next several years. However, fiber accessibility is likely to grow even further as a consequence of deployment activities by other cable operators, utilities, and municipalities.[194]

After a period of significant activity from 2000 through 2002, fiber deployment by cable operators fell off somewhat from 2003 to 2006. (Table 6, Chart 8) However, as offerings from competing providers achieve greater penetration, cable operators will likely respond with service improvements of their own, such as increased data speeds and more and better video offerings like expanded high definition television and video on demand.[195] To add the capacity necessary to provide these bandwidth-intensive applications, cable operators are expected to field almost 4.6 million miles of fiber between 2007 and 2010.[196] (Table 6)

In contrast to cable, fiber deployment by utilities and municipalities has undergone continued growth since 2003. (Chart 9) From 2007 to 2010 these entities are forecast to deploy almost 3.9 million miles of fiber as utilities ramp up activities associated with broadband over power lines (discussed below), and municipalities continue efforts to ensure their citizens have access to high-speed Internet access to their citizens.[197] (Table 6)

Broadband over Power Lines

As Table 6 and Chart 9 above illustrate, utilities and municipalities also contributed to the growth in fiber deployment and will likely continue to do so. Fiber deployment by utilities has been fueled in part by efforts to prepare to deliver broadband over power lines.[198] BPL offers the potential to leverage the nearly ubiquitous electricity infrastructure to provide broadband services to areas unserved by the more common wired and wireless technologies surveyed above. BPL uses the medium voltage (MV) and low voltage (LV) portions of the electrical grid to provide broadband Internet access to homes via radio frequency signals over the same facilities used to deliver electrical power.[199] BPL subscribers access the network using a modem plugged into a conventional electrical outlet.

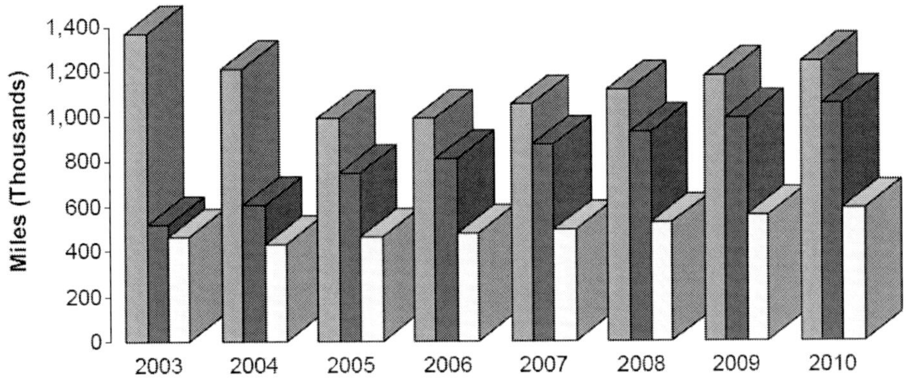

Source: NTIA (data from TIA Market Study, Table II-4.1).

Chart 9. Cable, Utility/Municipality, Private Fiber Deployment 2003-2010.

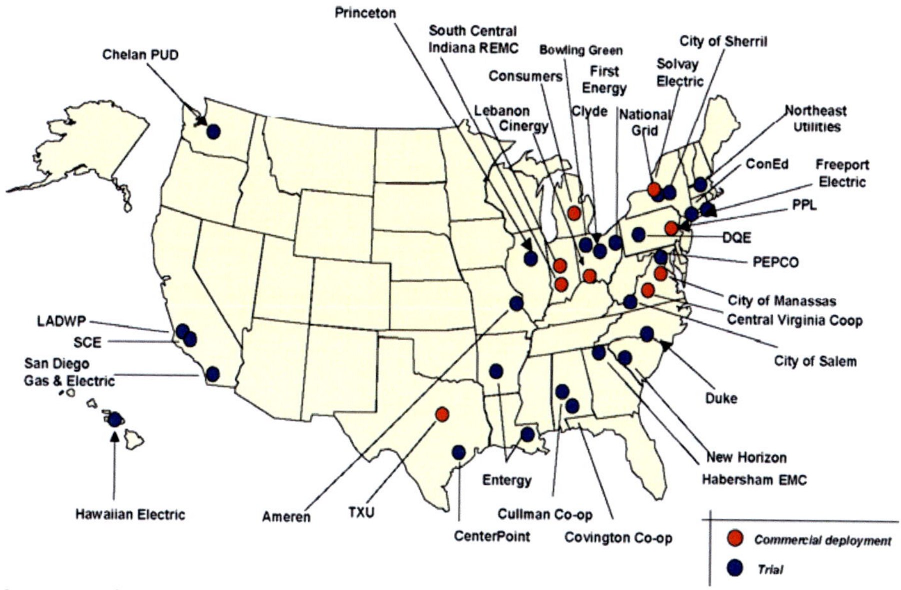

Source: United Power Line Council (updated as of July 10, 2007) (used with permission), http://uplc.utc.readyportal.net/file_depot/0-10000000/0-10000/7966/conman/2007+BPL+Update.pdf

Map 1.

While BPL has yet to make significant inroads in the broadband marketplace, it holds promise for the future. According to the United Power Line Council, there are about 35 BPL deployments in the United States, varying in size from pilot projects to large commercial developments.[200] (Map 1) The City of Manassas, Virginia has operated a BPL system in the city since 2005. Today, it provides 500- 800 Kbps Internet access to about 1,000 customers, although the city plans to increase speeds to 1-4 Mbps.[201] Duke Energy reportedly deploys BPL to about 50,000 homes in Cincinnati and neighboring portions of Kentucky.[202] The company has not released subscribership figures, but analysts estimate Duke Energy had some 7,500 customers as of October 2005.[203] TXU Electric Delivery has recently agreed to pay $150 million over 10 years to deploy BPL technology to about 2 million homes in the Dallas area.[204]

Reliable BPL subscribership figures are difficult to find. The FCC's most recent data identify fewer than 5,000 BPL customers as of year end 2006.[205] That figure appears low, however. TIA estimates 200,000 current BPL subscribers, increasing to 700,000 by 2010.[206] Another source forecast about 400,000 customers by the end of 2007, growing to 2.5 million by year end 2011.[207]

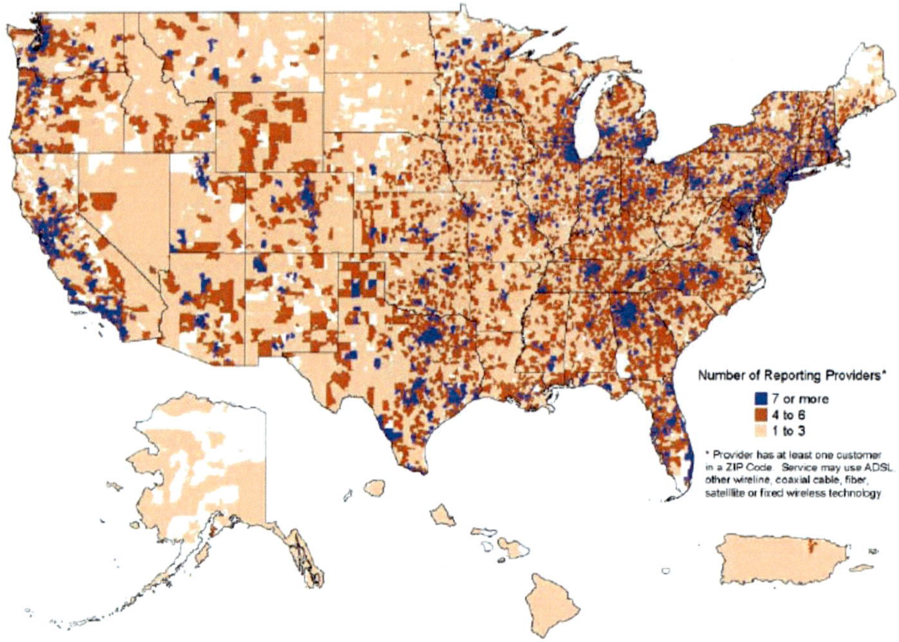

Map 2. High-Speed Providers by ZIP Code (As of December 31, 2003).

Some analysts suggest that BPL may have difficulty competing in areas where cable broadband or DSL is also available.[208] A manufacturer of BPL equipment indicates that the "sweet spot" for BPL may be communities with between 25,000 and 100,000 people, populous enough to yield economies of scale but not large enough to attract more than one or two other broadband providers.[209] BPL may also be an attractive option for ISPs seeking alternatives to cable and DSL. Thus, EarthLink, a prominent nonfacilities-based ISP, has invested in Current Communications, a major BPL equipment manufacturer, and has conducted BPL trials in several markets.[210] DirecTV recently announced plans to deliver Internet access via BPL in the Dallas-Fort Worth area.[211]

In summary, the statistical data outlined in the discussion above document the remarkable growth in broadband deployment that has occurred across all platforms over the last several years and the commensurate increase in adoption and usage of broadband technologies. The picture that emerges from those numbers becomes evident in the FCC's broadband coverage maps that appear on the following pages. A comparison of the maps from December 2003 (Map 2) and December 2006 (Map 3) graphically illustrates the geographic spread of broadband accessibility across the national landscape.

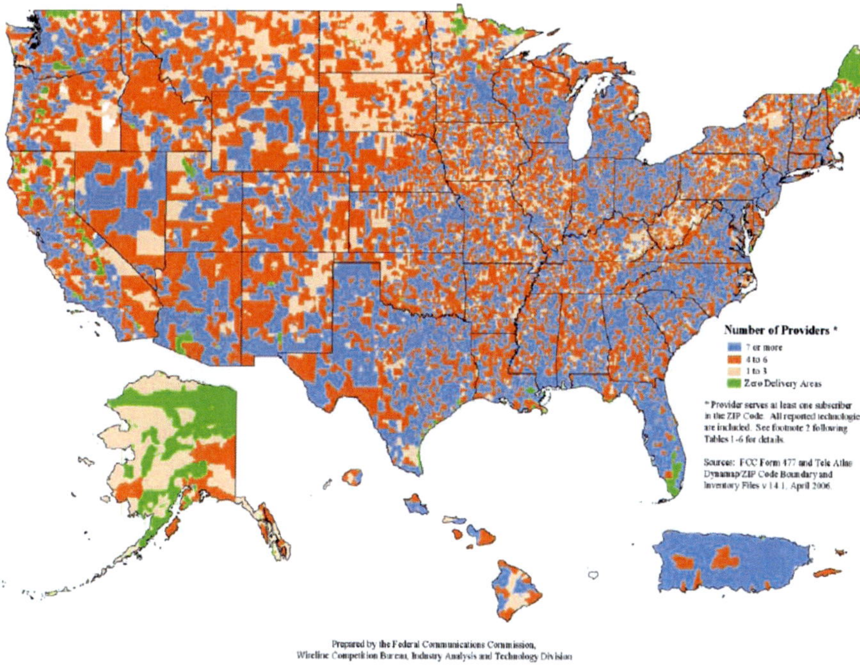

Map 3. High-Speed Providers by 5-Digit Geographical ZIP Code (As of December 31, 2006).

The December 2006 map – which does not account for growth over the last twelve months – depicts virtually ubiquitous territorial coverage of the United States by at least one broadband provider and makes evident the very high percentage of the nation served by at least four, and in many cases seven or more, providers. (Map 3)

Some gaps in territorial coverage continue to appear. However, a comparison of territorial broadband coverage (Map 3) relative to the nation's population density as reported by the United States Census Bureau (Map 4) and topography (Map 5) helps to place these gaps into proper context. Considering these factors, it appears clear, at least at the level of analytical granularity that the present data will support, that a vast majority of Americans (well over 90 percent) have access to broadband communications through one or more modality.[212]

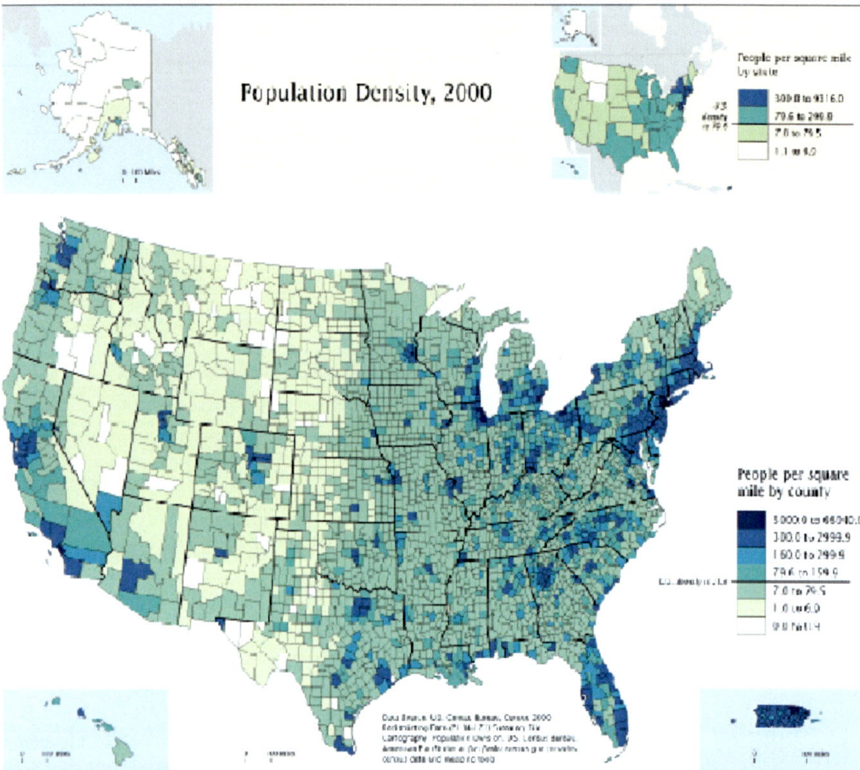

Map 4. Population Density, 2000.

It is also the case, however, that pockets of America exist where broadband service is not as robust, deployment has not been as rapid, and fewer choices are available. As the maps reflect, this is most likely the case in rural, mountainous, or otherwise remote areas of the country, where the population is most diffuse. Work remains to be done to identify these communities. To this end, improving the quality of our broadband penetration data will provide greater visibility into deployment patterns and specific challenges that need to be overcome.

The trend lines for investment in and build- out of new infrastructure and service capacity, discussed above and in the investment section that follows, provide a foundation for optimism. The surging competition that has so quickly brought broadband service to so much of the country can be expected to drive further deployment to presently underserved areas.

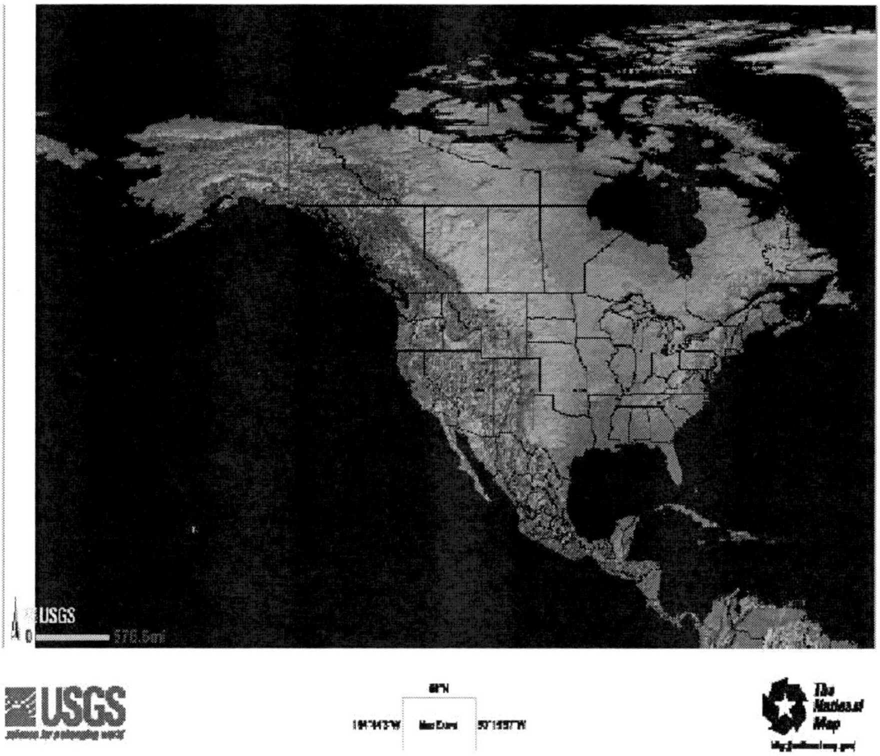

Map 5.

Robust Industry Investment – Creating Momentum for further Growth

The explosion of network development and deployment that has made broadband technology accessible to so many more Americans was not fortuitous; it stems directly from deliberate and massive increases in capital investment made by service providers across all subsectors of the broadband marketplace. The previous section discussed many of the purposes for which this capital was used. The figures below document how much was spent and the economic impact of this investment in real dollars. More importantly, analysts' forecasts, based on trend patterns and market drivers, provide some indication of expected future investment and the direction in which the broadband services market is moving: toward accelerating growth and further accessibility. TIA forecasts that spending to support broadband network infrastructure will rise over the next four years from $15.2 billion in 2007 to $23 billion in 2010. (Table 8, *infra.*)

Since 2003, overall industry investment in infrastructure and equipment has increased significantly. According to TIA, spending by carriers on telecommunications equipment, including central office, fiber, asynchronous transfer mode (ATM), and frame relay equipment, rose from $15.2 billion in 2003 to $24.4 billion in 2007. (Table 7) Verizon reports that "[i]n the three years since federal regulators began dismantling network sharing and pricing regulation of broadband networks, Verizon's total capital expenditures were more than $45 billion, including $12.8 billion in 2004, $15.0 billion in 2005, and $17.1 billion in 2006."[213] Other segments of the broadband market have exhibited similarly strong investment patterns.

Table 7. Spending by Carriers on Telecommunications Equipment by Category in the United States ($ Millions)

Year	Central Office Equipment	Fiber	ATM	Frame Relay	Total
2000	27,000	20,100	2,690	1,815	51,605
2001	26,600	14,100	2,050	1,800	44,550
2002	12,000	5,270	1,150	1,780	20,200
2003	8,100	4,235	1,100	1,750	15,185
2004	7,500	6,840	1,130	1,730	17,200
2005	7,650	9,640	1,075	1,710	20,075
2006	9,400	10,460	1,000	1,650	22,510
2007	10,600	11,320	925	1,575	24,420
2008	11,700	12,310	850	1,500	26,360
2009	12,700	12,970	750	1,400	27,820
2010	13,500	13,480	650	1,300	28,930

Source: TIA Market Study, Table II-4.4 (*FCC, TIA, Wilkofsky Gruen Assoc.*) (used with permission).

High-Speed Cable

From 1996 through 2006, according to NCTA, cable companies spent more than $117 billion in capital expenditures to develop their broadband networks.[214] (Chart 10, *infra*) The industry reports that it has invested $23 billion in the last two years alone to enhance and improve its advanced, interactive, hybrid-fiber coaxial network."[215] During 2007, cable operators were expected to spend another $13.7 billion on infrastructure – a 10 percent increase over 2006 – continuing an upward trend.[216]

There is reason to expect cable investment to increase in the future. TIA's analysis predicts that spending on services in support of cable network infrastructure equipment will almost double from $7.8 billion in 2003 to $15.2 billion by 2010. (Table 8) TIA notes that, although the present capacity of most cable systems is sufficient, the expected increase in customer demand for high-definition television channels and video-on-demand (VOD) services will require cable operators to expand their infrastructure to deliver those services.[217] Accordingly, TIA projects significant growth in fiber deployment by cable operators from 2007 through 2010 with a cumulative additional deployment during that period of 4.6 million miles.[218]

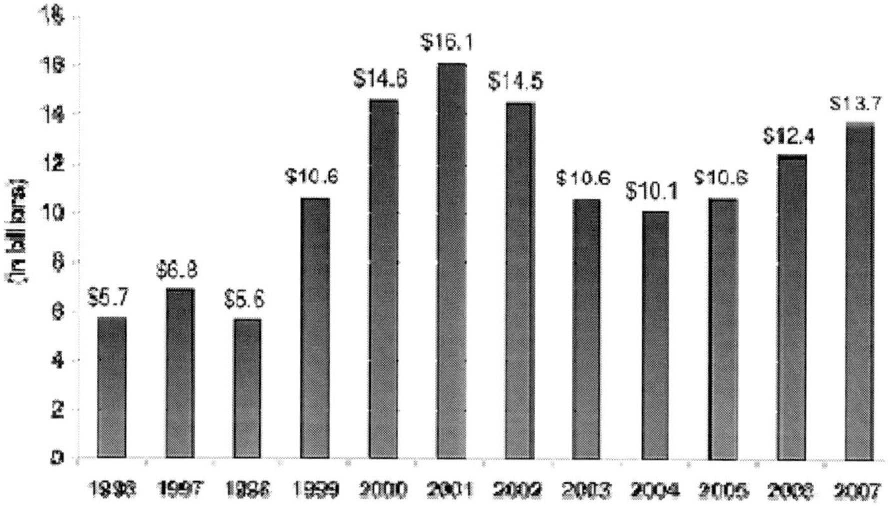

Source: NCTA (SNL Kagan) (used with permission)
http://www.ncta.com/Statistic/Statistic/InfrastructureExpenditures.aspx

Chart 10. Cable Industry Infrastructure Expenditures.

Fixed and Mobile Wireless

As noted above, broadband-fueled growth in the fixed and mobile wireless services markets has been especially strong over the past few years, and capital investment in these sectors has been correspondingly vigorous. According to data from TIA and CTIA, factors such as the AWS auction, deployment of 3G networks, and the build-out of additional infrastructure needed to support additional capacity for new, advanced services helped spur U.S. wireless providers to increase capital investment spending from $18.9 billion in 2003 to $30 billion in 2007.[219] (*See* Chart 11, *infra*)

As these factors persist, and new developments such as WiMAX deployment emerge, TIA expects this level of investment to continue with carriers expected to spend nearly $100 billion over the next three years.[220] MuniWireless LLC, predicted 2007 investment in municipal wireless systems in the United States to rise 35 percent, surpassing $329 million,[221] and TIA expects revenues for WiMAX infrastructure to increase from $175 million in 2006 to $4.4 billion in 2008.[222]

Fiber

Table 7 reflects the investment trends with respect to fiber-related telecommunications equipment. As the table indicates, after an intensive period of investment and deployment at the outset of the decade, fiber spending decreased somewhat through 2003 but has been on an upward trend since then. TIA estimates that spending on fiber equipment will reach $11.3 billion in 2007 and will grow to almost $13.5 billion annually by 2010, representing a total investment in excess of $50 billion over the period. (*See* Table 7)

As discussed above, in addition to actions by a host of other companies (including smaller telephone companies, cable operators, utilities and municipalities), the Verizon FiOS and AT&T U-verse platforms will likely drive significant investment. Verizon is planning to spend $23 billion to make fiber available to 18 million homes by the end of 2010.[223] Similarly, in its effort to reach 17 million households by the end of 2008, AT&T plans to spend between $4.5 billion and $5.0 billion on U-verse, to deliver broadband, IP-enabled high-definition video, ultra-high speed Internet, and voice services.[224] The third major ILEC, Qwest Communications International, has also stated that it plans to spend $300 million over the next year to deploy fiber upgrades to its network to deliver 20 Mbps connectivity to 1.5 million more homes.[225]

Table 8. Spending on Services in Support of Network Infrastructure Equipment in the United States ($ Millions)

Year	POTS (Plain Old Telephone Service)	Cable	Broadband	Total
1999	19,200	4,800	1,000	25,000
2000	21,800	5,800	3,200	30,800
2001	24,200	6,900	7,000	38,100
2002	23,900	8,300	9,700	41,900
2003	22,800	7,800	8,700	39,300
2004	24,900	8,800	10,500	44,200
2005	27,400	10,200	12,500	50,100
2006	30,300	11,700	14,700	56,700
2007	32,900	13,000	15,200	61,100
2008	32,200	13,800	18,000	67,000
2009	37,300	14,500	20,700	72,500
2010	39,500	15,200	23,000	77,700

Source: TIA Market Study, Table II-4.6 (D.F. Blumberg & Assoc., TIA, Wilkofsky Gruen Assoc.) (used with permission).

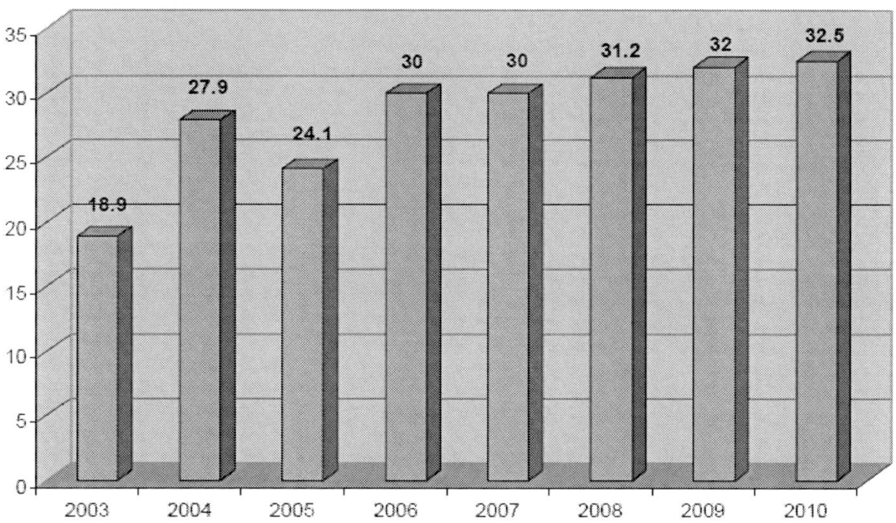

Source: TIA Market Study, Figure IV–1 .6 (TIA; CTIA – The Wireless Association; Wilkofsky Gruen Associates).

Chart 11. Wireless Capital Expenditure in the United States ($ Billions).

These capital investment figures, and the trends they reflect, demonstrate the confidence that exists – on the part of service providers, equipment manufacturers, application developers, and, perhaps more importantly, on the part of the stockholders and financial institutions that fund them – in the strength and potential of the U.S. broadband marketplace. These investment initiatives will help to ensure that the accomplishments of the last four years in securing broadband access for all Americans do not languish, but instead contribute to continued growth in the future.

AFFORDABILITY: COMPETITION DRIVING DOWN PRICES

Escalating competition among broadband platforms and service providers, coupled with sustained business investment, has clearly yielded a proliferation of innovative new communications and entertainment services for consumers. Less evident, but equally important, this competition is also making those services more affordable.

As the FTC reported in its broadband competition study:

> Prices for DSL broadband services have also fallen rapidly as the telephone companies have competed aggressively to take market share from the cable companies. By one estimate, the average monthly revenue per user of DSL service decreased from $ 40 in 2002 to $ 31 in 2006. From May 2005 to April 2006, AT&T reduced the monthly price of 3.0 Mbps DSL service from $29.95 to $1 7.99."[226]

Moreover, analysts have noted that competition from DSL has also exerted price pressure on high-speed cable services. The FTC noted that, adjusting for quality of service, the price of cable broadband has fallen.[227] TIA has tracked a similar trend, noting steady decreases in the average price of high-speed cable access since 2004 corresponding to lower average prices for DSL service. (Chart 12)

TIA expects competition between cable operators and telephone companies to continue to exert downward pressure on cable modem rates, forecasting that the average monthly bill for cable modem service to drop to $38.49 by 2010.[228] FTTH deployments are also affecting cable prices. Verizon, citing research conducted by Bank of America, notes that cable operators in FiOS markets have lowered their prices or increased the quality of service packages in response to the competition.[229]

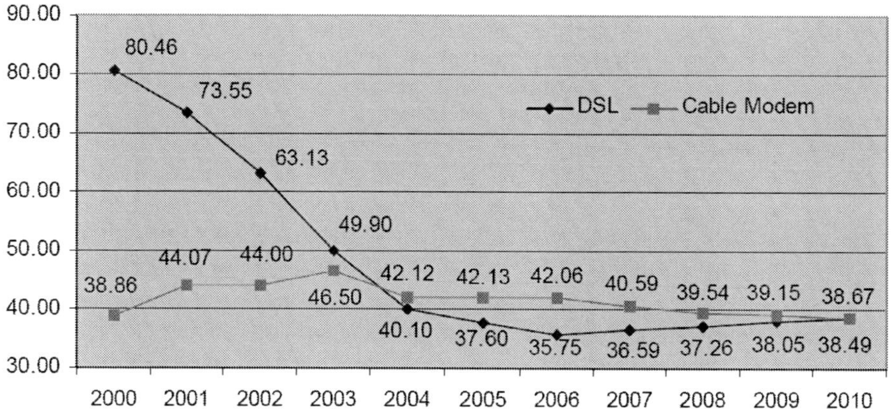

Source: TIA Market Study, Figure II–2.17 (In-Stat; TIA; Wilkofsky Gruen Associates).

Chart 12. Average Monthly Broadband Fees in the United States ($).

Pricing data from CTIA as shown in Table 3 above reflect that average local monthly bills for the wireless market have remained relatively stable, rising on average less than a dime per year from $49.46 in 2003 to $49.94 in 2007. However, during the same interval, overall usage grew from just under 400 billion minutes in June 2003 to almost two trillion minutes by June of 2007, with a staggering increase of over 200 billion minutes in the last year alone. (Chart 13) The relative stability of prices juxtaposed against this dramatic increase in usage demonstrates a significant decrease in relative costs for consumers.

Table 5 above reflects that satellite broadband, although ubiquitous, remains among the more expensive of the broadband service options available to consumers when compared against the relative costs and transmission speeds offered by competing terrestrial platforms. A significant factor contributing to this disparity is the cost associated with the purchase and installation of the customer premises equipment (CPE) required to receive the satellite broadband service.[230]

However, even in the satellite sector, prices are coming down. For example, WildBlue last year reduced the price for professional installation by over 55 percent, from $179.95 to $79.95.[231] The industry expects that prices will begin to fall more quickly when satellite providers reach a "critical mass" of subscribers and begin to deploy new systems such as the hybrid satellite/ATC networks discussed above.[232] Citing the example of direct-tohome satellite television providers, DirecTV and DISH Network, SIA indicates that these developments will enable satellite providers to leverage economies of scale that will enable them fully to subsidize their CPE or, at least, price it in a range more typical for other consumer electronics equipment. [233]

HOW DOES BROADBAND PRICING COMPARE TO OTHER BUNDLES OF GOODS AND SERVICES?

From 2002 through 2007, the nation's Consumer Price Index (CPI) increased from 179.9 to 205.7 (first half of 2007). U.S. Department of Labor, Bureau of Labor Statistics, Branch of Economic Analysis and Information, Consumer Price Index, Historical Data Tables for the CPI-U and CPI-W, All Items (1982-84=100), Summary of Annual and Semi-annual Indexes (http://www.bls.gov/ro3/fax 91 25jdf).

The CPI is a measure of the average change over time in the prices paid by urban consumers for a market basket of consumer goods and services. The broadest, most comprehensive CPI is the

Consumer Price Index for All Urban Consumers (CPI-U) for the U.S. City Average for All Items, 1982-84=100.

While pricing for broadband decreased or stabilized, pricing for many other basic goods and services has increased. For example, CPI data from 2002 to 2005 show the following price trends in other major consumer cost centers:

FOOD

up 10.3 percent

FOOD (AWAY FROM HOME)

up 15.7 percent

TRANSPORTATION

up 7.5 percent

ENTERTAINMENT

*up*14.9 percent

EDUCATION

up 25.0 percent

Source: U.S. Department of Labor, Bureau of Labor Statistics, Consumer Expenditures Annual Reports, 2004 (http://www.bls.gov/cex/csxann04jdf) and 2005 (http://www.bls.gov/cex/csxann05jdf).

Providers are working aggressively to achieve this critical mass. WildBlue has established partnerships with the NRTC, AT&T, DirecTV, and DISH Network to distribute its service.[234] Similarly, Hughes recently announced an agreement with Wal-Mart for the low-price retailer to resell HughesNet broadband services in 800 of its stores.[235] Noting that Wal-Mart's entry into a new market often "tends to push down prices," some analysts believe that the arrangement with Hughes could similarly affect other broadband service providers.[236]

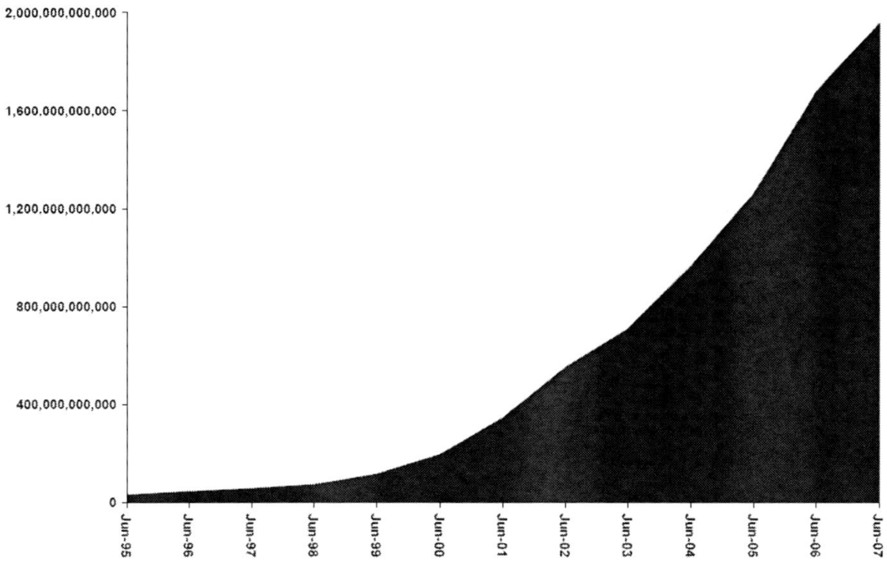

Source: CTIA, *Wireless Industry Indices*, Midyear 2007 Results (November 30, 2007), Section 6.3 "MOU Trend Analysis," Table 85 (used with permission).

Chart 13. Annualized Minutes of Use (July 1- June 30).

Satellite providers are also exploring other pricing models to make their services more affordable to consumers. For example, Hughes offers residential subscribers a pricing option that rolls the up-front charges into the monthly service fee; thus, instead of paying $299.98 for equipment and installation (after rebate) plus the $59.99 monthly charge for basic HughesNet Home Service (or $69.99 or $79.99 for HughesNet Professional or Professional Plus, respectively) (*see* Table 5), the customer can pay $20.00 more per month (*i.e.*, $79.99/$89.99/$99.99) for 24 months and eliminate the up-front charge.[237]

This option requires a 24-month service commitment (after which the monthly fee returns to the standard amount), but customers also receive an additional $100 mail-in rebate.[238] While the resulting service fee is still appreciably higher than those for comparable terrestrial services, this type of pricing approach helps to mitigate the disincentive to subscribership that presently exists and should help to stimulate consumer uptake.

Accelerating technological evolution and the emergence of innovative and attractive new applications and services are kindling consumer interest and stoking demand for broadband services across multiple platforms. That strong demand, coupled with a proliferation of service providers seeking to satisfy it, is generating competition that is producing lower prices and more choices for consumers in all sectors.

Chapter 4

CONCLUSION

The swift growth in broadband technologies and services that has occurred over the last several years – and the correspondingly rapid economic and cultural changes that have attended it – underscore the importance of broadband communications as a transformative resource for the nation. The capacity to manage large amounts of information (whether voice/audio, data, or video) and to quickly and efficiently exchange it with others down the street or across the globe is no longer a luxury but, like the telephone over a century ago, is quickly becoming an essential tool for life and commerce in the modern world. The same technology that enables governments, banks, corporations and other institutions to manage affairs of state and international finance and trade now makes critical contributions to health care, education, public safety, and the productivity of individuals and small businesses, as well as a host of other activities. For this reason, the President has made it a priority to ensure that all Americans have affordable access to this important resource by harnessing the power of the competitive marketplace.

As this report demonstrates, a reasonable assessment of the available data indicates that the nation has, to a very great degree, realized this objective. The FCC reports that, as of the end of 2006, over 99 percent of all U.S. zip codes received broadband service from at least one provider, and those zip codes encompass over 99 percent of the nation's population.[239] In addition, recent Census Bureau data and the individual growth patterns in each of the broadband modal sectors, documented above, make it likely that broadband accessibility, penetration and usage have grown further since the Commission collected its information and are continuing to do so. Finally, pricing and usage information for the major broadband sectors also indicate that these services are becoming more affordable for consumers as time passes.

Nevertheless, it is also clear that more work remains to be done both by the government and by the private sector to improve broadband data. The FCC's most current broadband coverage map continues to show areas of potential service gaps, albeit primarily in sparsely populated remote areas. Anecdotal reports also contain stories of individuals who report being unable to access a broadband connection for one reason or another. In this regard, the general nature of the present data necessarily makes it difficult to identify precisely the specific characteristics of broadband service available in particular areas. The U.S. Government is taking action to improve its data collection tools to obtain more granular information that will provide a more detailed view into these issues.

The recent history of broadband deployment in the United States stands as a strong endorsement for the principles embodied in the 1996 Telecommunications Act. The remarkable array of broadband access technologies, applications, and services that our nation enjoys today flows from past ingenuity and capital investment, but it is also carries the momentum that will further propagate these technologies and services and stimulate investment in the future. Copper and coaxial cable are yielding to optical fiber, and even as deployment of 3G wireless technologies proceeds, planning is already underway for 4G systems. Each new roll-out of a technology or service increases adoption, which fuels demand among other consumers, and which, in turn, provides the business case for further investment to meet demand and provision for future growth.

This cycle is the natural outgrowth of fervent competition embraced in the Telecommunications Act of 1996 and cultivated through effective technology, regulatory, and fiscal policies. The evidence is clear: The pro- competition telecommunications policies underpinning the Administration's broadband strategy has stabilized the market for planning and investment, provided needed spectrum resources to expand existing services and support new innovations, and spurred the development and deployment of technologies and infrastructure necessary to deliver them. These policies have delivered outstanding results and hold great promise for the future.

Endnotes

1 R. Jaques, *Doctors Perform Surgery Over the Web*, VNUnet.com (Aug. 9, 2005), http://www.vnunet.com/articles/print/2140922 (last visited Jan. 15, 2008).

2 Associated Press, *Soldiers Abroad to Watch Graduations Online* (May 18, 2006), http://www.msnbc.msn.com/id/12861456/ (last visited Jan. 15, 2008).

3 R. Callimachi, *Biggest Wi-Fi Cloud is in Rural Oregon* (Oct. 16, 2005), *reprinted in* USA Today, http://www.usatoday.com/tech/products/services/2005-10-16-oregon-wi-fix.htm (last visited Jan. 15, 2008).

4 *Dallas Launches Citywide Interoperability Network*, Government Technology (Nov. 16, 2006), http://www.govtech.com/gt/articles/102406 (last visited Jan. 15, 2008).

5 B. Ipsen, *Last FrontierLlinks with First Frontier*, First Alaskans (Jan./Feb. 2007), at 23-24.

6 Telecommunications Act of 1996, Pub. L. No. 104-104, 110 Stat. 56 (1996) (*amending* the Communications Act of 1934, 47 U.S.C. §§ 1-714).

7 *Id.* § 509, *codified at* 47 U.S.C. § 230(b) (1)-(2).

8 *Promoting Innovation and Competitiveness: President Bush's Technology Agenda*, the White House (June 2002), http://www.gcrio.org/OnLnDoc/pdf/technologyagenda.pdf.

9 *Id.* at 1.

10 American Competitiveness Initiative (Feb. 2, 2006), the White House, http://www.whitehouse.gov/stateoftheunion/2006/aci/index.html.

11 *Id.* at 3.

12 *Id.*

13 Memorandum for the Heads of Executive Departments and Agencies, Subject: Spectrum Policy for the 21st Century, 39 Weekly Comp. Pres. Doc. 726, 727 (May 23, 2003), http://www.whitehouse.gov/news/releases/2003/06/20030605-4.html (President's Spectrum Policy Memo). *See also* Presidential Memorandum on Spectrum Policy for the 21st Century, 69 Fed. Reg. 1569 (Jan. 9, 2004).

14 *Spectrum Management for the 21st Century: Plan to Implement Recommendations of the President's Spectrum Policy Initiative*, U.S. Department of Commerce, NTIA (2006), http://www.doc.gov/osmhome/reports/ImplementationPlan2006.htm; Transcript of Hearings, *Improving Spectrum Management through Economic and other Incentives* (Feb. 28, 2006), http://www.ntia.doc.gov/forums/2006/specman/

spectrumworkshop022806.pdf, and (March 1, 2006), http://www.ntia.doc.gov/forums/2006/specman/spectrumworkshop022806.pdf.

15 D. Jorgenson, *The Telecommunications Challenge: Changing Technologies and Evolving Policies,* Concluding Remarks at 158 (Charles W. Wessner, ed., The National Academies Press, 2006).

16 E. J. Rutkowski, *3GSM 2007: Video Review of I-mate Ultimate 5150* (Feb. 13, 2007), http://msmobiles.com/news.php/6059.html (last visited Jan. 15, 2008); tgdaily, *Broadcom Announces 7.2 Mb/s Single-chip HSDPA Mobile Phone Processor*, (Apr. 6, 2006), http://www.tgdaily.com/content/view/25579/107/ (last visited Jan. 15, 2008).

17 Cellular-News, *Mobile-phone Premium Content Market to Reach $40 Billion by 2010*, (March 8, 2006), http://www.cellularnews.com/story/16425.php (market projection by iSuppli Corp.) (last visited Jan. 15, 2008).

18 Telecommunications Industry Association (TIA), *2007 Telecommunications Market Review and Forecast*, at 167 and Table IV-2.2 (TIA Market Study).

19 Commercial Spectrum Enhancement Act, Pub. L. No. 108-494, 118 Stat. 3991-97 (2004), *codified at* 47 U.S.C. §§ 309(j), 923(g), 928 (2006).

20 Public Notice, *Auction of Advanced Wireless Services Licenses Closes; Winning Bidders Announced for Auction No. 66*, FCC DA 06-1882 (Sept. 20, 2006) (noting that a total of $13,700,267,150 in net bids had been raised in the AWS auction).

21 *Id.*

22 Statement of Administration Policy, H.R. 4241 – the Deficit Reduction Act of 2005 (Nov. 17, 2005) at 3, http://www.whitehouse.gov/omb/legislative/sap/109-1/hr4241sap-h.pdf; Statement of Administration Policy, S.1932 – the Deficit Reduction and Omnibus Reconciliation Act of 2005 (Nov. 1, 2005), http://www.whitehouse.gov/omb/legislative/sap/109-1/s1932sap-s.pdf at 3.

23 Comments of the National Telecommunications and Information Administration in ET Doc. No. 03-122, *Revision of Parts 2 and 15 of the Commission's Rules to Permit Unlicensed National Information Infrastructure (U-NII) Devices in the 5 GHz Band*, (Oct. 1, 2003), http://www.ntia.doc.gov/ntiahome/fccfilings/2003/et03-122comments10012003.pdf.

24 Comments of the National Telecommunications and Information Administration in ET Docket No. 98-153, *Revision of Part 15 of the Commission's Rules Regarding Ultra-wideband Transmission Systems* (Jan. 15, 2004), http://www.ntia.doc.gov/ntiahome/fccfilings/2004/ETDocket98-153comments01152004.pdf; Comments of the National Telecommunications

and Information Administration in ET Docket No. 98-153, *Revision of Part 15 of the Commission's Rules Regarding Ultrawideband Transmission Systems* (Sept. 12, 2000), http://www.ntia.doc.gov/ntiahome/fccfilings/2000/et98153comments.htm.

25 News Release, *NTIA and FCC Launch On-line Registration for High-Speed Wireless Links Sharing Spectrum in the 70-80-90 GHz Bands,* U.S. Department of Commerce (Feb. 8, 2005), http://www.ntia.doc.gov/ntiahome/press/2005/708090website02082005.htm.

26 *Amendment of Part 15 Regarding New Requirements and Measurement Guidelines for Access Broadband over Power Line Systems*, ET Doc. No. 04-37, Report and Order, 19 FCC Rcd 21265 (2004), recon., 21 FCC Rcd 9308 (2006); *see also* News Release, *NTIA Report Lays Groundwork for Responsible Deployment of Broadband Over Power Lines (BPL)*, NTIA (April 27, 2004), http://www.ntia.doc.gov/ntiahome/press/2004/bplrelease04272004.pdf.

27 From spring 2003 through summer 2007, NTIA conducted a series of field tests and performed extensive analysis relative to the potential affects of BPL systems on Federal radiocommunications systems operating in the 1.7 to 80 MHz band. The results of this analysis are documented in NTIA's BPL Phase 1 Study Report, *Potential Interference from Broadband Over Power Line (BPL) Systems to Federal Government Radiocommunications at 1.7-80 MHz: Phase 1 Study,* NTIA Report 04-413, U.S. Dept. of Commerce, NTIA (2004), http://www.ntia.doc.gov/ntiahome/fccfilings/2004/bpl/index.html, and its Phase 2 Study Report, *Potential Interference From Broadband Over Power Line (BPL) Systems To Federal Government Radiocommunication Systems at 1.7 - 80 MHz: Phase 2 Study*, NTIA Report No. 08-450, U.S. Dept. of Commerce, NTIA (2004), http://www.ntia.doc.gov/osmhome/reports/2007/bpl2007.html . They were submitted to the FCC along with NTIA's Comments on the BPL Notice of Proposed Rulemaking.

28 *See e.g.*, News Release, *Assistant Secretary of Commerce Nancy J. Victory Commends FCC for Initiating Broadband Proceeding*, NTIA (Feb. 14, 2002) (commending the FCC for initiating *Appropriate Framework for Broadband Access to the Internet over Wireline Facilities*, CC Doc. No. 02-33, Notice of Proposed Rulemaking, 17 FCC Rcd 3019 (2002)), http://www.ntia.doc.gov/ntiahome/press/2002/njv21402.htm.

29 *Review of the Section 251 Unbundling Obligations of Incumbent Local Exchange Carriers; Implementation of the Local Competition Provisions of the Telecommunications Act of 1996,* CC Doc. No. 01-338, Order on Reconsideration, 19 FCC Rcd 15856 (2004) (fiber loops delivered to

residential multiple dwelling units); *Review of the Section 251 Unbundling Obligations of Incumbent Local Exchange Carriers,* CC Doc. No. 01-338, Order on Reconsideration, 19 FCC Rcd 20293 (2004) (fiber to the curb); FCC Triennial Review Order, 18 FCC Rcd 16978 (fiber to the home loops). *See also* News Release, *NTIA Acts on President's Call to Clear Out 'Underbrush of Regulation' Towards Broadband Deployment: Urges FCC to Clarify Definitions in "Fiber-to-the-Home" Rules for Residential "Multi-Unit Premises*," NTIA, (April 29, 2004), http://www.ntia.doc.gov/ntiahome/press/2004/mduletter04292004.htm; News Release, *Statement by Assistant Secretary Michael D. Gallagher on the FCC's Order on Reconsideration of the Triennial Review Order*, NTIA (Aug. 4, 2004) ("The Commission's action today creates the regulatory certainty necessary to speed the deployment of broadband to millions of Americans living in apartment buildings, and brings us closer to meeting the President's goal of universal and affordable broadband access by 2007."), http://www.ntia.doc.gov/ntiahome/press/2004/tro08042004.html.

30 FCC Triennial Review Order, 18 FCC Rcd 17126, ¶ 243, 17132-17141.

31 *Inquiry Concerning High-Speed Access to the Internet Over Cable and Other Facilities; Internet Over Cable*, GN Doc. No. 00-185, Declaratory Ruling and Notice of Proposed Rulemaking, 17 FCC Rcd 4798 (2002) (Cable Modem Declaratory Ruling), *aff'd sub nom., National Cable & Telecommunications Ass'n v. Brand X Internet Servs.*, 545 U.S. 967 (2005). In making this determination, the Commission indicated that its primary policy goal was to "encourage the ubiquitous availability of broadband to all Americans" and the belief that "broadband services should exist in a minimal regulatory environment that promotes investment and innovation in a competitive market." Cable Modem Declaratory Ruling, 17 FCC Rcd at 4801-4802.

32 *Appropriate Framework for Broadband Access to the Internet over Wireline Facilities*, CC Doc. No. 02-33, Report and Order and Notice of Proposed Rulemaking, 20 FCC Rcd 14853 (2005) (Broadband Access Order), *aff'd sub nom., Time Warner Telecommunications v. FCC*, 507 F.3d 205 (3rd Cir. 2007). *See also* News Release, *Statement by Assistant Secretary Michael D. Gallagher on the FCC's Adoption of Wireline Broadband Internet Access Order*, NTIA (Aug. 5, 2005), http://www.ntia.doc.gov/ntiahome/press/2005/gallagher08052005.htm.

33 *Broadband Access Order*, 20 FCC Rcd at 14975.

34 *United Power Line Council's Petition for Declaratory Ruling Regarding the Classification of Broadband over Power Line Internet Access Service as an*

Information Service, WC Doc. No. 06-10, Memorandum Opinion and Order, 21 FCC Rcd 13281 (2006).

35 *Appropriate Regulatory Treatment for Broadband Access to the Internet Over Wireless Networks*, WT Doc. No. 07-53, Declaratory Ruling, 22 FCC Rcd 5901 (2007).

36 *See* News Release, *Acting Assistant Secretary of Commerce Michael D. Gallagher Commends FCC For Favoring "Innovation, Not Regulation,"* NTIA (Feb. 12, 2004), http://www.ntia.doc.gov/ntiahome/press/2004/voip02122004.htm (Gallagher February 12 Statement). Assistant Secretary Gallagher referred to the following FCC decisions: *IP-Enabled Services,* WC Doc. No. 04-36, Notice of Proposed Rulemaking, 19 FCC Rcd 4863 (2004); and *Petition for Declaratory Ruling That Pulver.com's Free World Dialup is Neither Telecommunications Nor a Telecommunications Service*, WC Doc. No. 03-45, Memorandum Opinion and Order, 19 FCC Rcd 3307 (2004).

37 *See* News Release, *Statement by Assistant Secretary Michael D. Gallagher on FCC's VOIP Action,* NTIA (Nov. 9, 2004), http://www.ntia.doc.gov/ntiahome/press/2004/voip11092004.html (referring to *Vonage Holdings Corporation Petition for Declaratory Ruling Concerning an Order of the Minnesota Public Utilities Commission*, WC Doc. No. 03-211, Memorandum Opinion and Order, 19 FCC Rcd 22404 (2004), *aff'd sub nom. Minnesota Pub. Utils Comm'n v. FCC*, 483 F.3d 570 (8th Cir. 2007). Consistent with the over-arching principles identified above, however, the Administration has also concurred with the Commission's requirement that VoIP providers include emergency calling capabilities (E911) for their subscribers and afford law enforcement authorities access to their networks as appropriate to ensure public safety. *See* Statement of Administration Policy, H.R. 5252 – Communications Opportunity, Promotion, and Enhancement Act of 2006 (June 8, 2006) (E911); Gallagher February 12 Statement, *supra* note 36 (law enforcement access).

38 Statement of Administration Policy, H.R. 5252 – the Communications Opportunity, Promotion, and Enhancement Act of 2006 (June 8, 2006) at 1, http://www.whitehouse.gov/omb/legislative/sap/109-2/hr5252sap-h.pdf.

39 *Implementation of Section 621(a)(1) of the Cable Communications Policy Act of 1984 as amended by the Cable Television Consumer Protection and Competition Act of 1992*, MB Docket No. 05-311, Report and Order and Further Notice of Proposed Rulemaking, 22 FCC Rcd 5101, 5133 (2006).

40 *Id.* at 5103.

41 *Id.*

42 Memorandum for the Heads of Executive Departments and Agencies, *Improving Rights-of-Way Management Across Federal Lands to Spur Greater Broadband Deployment*, 40 Weekly Comp. Pres. Doc. 696 (April 26, 2004), http://www.whitehouse.gov/news/releases/2004/04/20040426-2.html (Broadband Rights-of-Way Memorandum).

43 *Id.*

44 News Release, *President Signs Internet Access Taxation Moratorium*, http://www.whitehouse.gov/news/releases/2001/11/20011129.html; Statement of Administration Policy, S. 150 - Internet Tax Non-discrimination Act of 2003 (Nov. 6, 2003), http://www.whitehouse.gov/omb/legislative/sap/108-1/s150sap-s.pdf; Statement of Administration Policy, H.R. 49 - Internet Tax Nondiscrimination Act (Sept. 17, 2003), http://www.whitehouse. gov/omb/legislative/sap/108-1/hr49sap-h.pdf; News Release, *President George W. Bush Signs S.150, the Internet Tax Nondiscrimination Act*, http:// www.whitehouse.gov/news/releases/2004/12/images/20041203-6120304 taxthjpg-515h.html; News Release, *President George W. Bush Signs H.R. 3678, the Internet Tax Freedom Act Amendments Act of 2007*, the White House, http://www.whitehouse.gov/news/releases/2007/10/images/20071031-17d-0350-2-515h.html.

45 Jobs and Growth Tax Relief Reconciliation Tax Relief Act of 2003, Pub. L. No. 108-27, 117 Stat. 752 (2003), http://frwebgate.access.gpo.gov/cgi-bin/getdoc.cgi?dbname=108congpubliclaws&docid=f:publ027.108.pdf; and Statement of Administration Policy, S. 1054 – Jobs and Growth Tax Act of 2003 (May 14, 2003), http://www.whitehouse.gov/omb/legislative/sap/108-1/s1054sap-s.pdf. *See also* Press Release, *President Signs Jobs and Growth Tax Relief Reconciliation Act of 2003*, the White House (May 28, 2003), http://www.whitehouse.gov/news/releases/2003/05/20030528- 9.html.

46 Tax Relief and Health Care Act of 2006, Pub. L. No. 109-432, 120 Stat. 2922, 2934-36 (2006) (extending the R&E tax credit from December 31, 2005 to December 31, 2007), http://frwebgate.access.gpo.gov/cgibin/getdoc. cgi?dbname= 109 cong public laws&docid=f:publ432.109.pdf. See also News Release, *Fact Sheet: Tax Relief and Health Care Act of 2006*, the White House (Dec. 20, 2006), http://www.whitehouse.gov/news/releases/2006/12/ 20061220.html.

47 *See* American Competitiveness Initiative, *supra* note 10; *see also* News Release, *President Bush Discusses American Competitiveness Initiative During Press Conference* (Aug. 9, 2007), the White House, http://www. whitehouse.gov/news/releases/2007/08/20070809- 1.html.

48 *Broadband Investment Report,* USDA (2007), http://www.usda.gov/rus/telecom/broadband/pdf/BroadbandInvestmentReport.pdf.

49 *USDA Rural Development: Bringing Broadband to Rural America*, USDA (June 2007) at 2, 4, http://www.rurdev.usda.gov/rd/pubs/RDBroadbandRpt.pdf.

50 *Broadband Investment Report*, *supra* note 48 (as mandated by the Rural Electrification Loan Restructuring Act of 1993, all facilities financed under the RE Act of 1936 must be capable of delivering 1 Mb/s service).

51 *Id.*

52 USDATelecommunications Program Home Page, http://www.usda.gov/rus/telecom/ (last visited Jan. 15, 2008).

53 *Broadband Investment Report, supra* note 48.

54 *Community Connect Broadband Program, Grant Application Guide,* USDA, Rural Development Utilities Programs (2007) at 2, http://www.usda.gov/rus/telecom/commconnect/2007programinformation/2007-application-guide-7-5-07.pdf. (last visited Jan. 15, 2008).

55 *Id.*

56 News Release, *Veneman Announces $11.3 Million in Broadband Technology Grant*, USDA, Release No. 0328.03 (Sept. 24, 2003), http://www.usda.gov/documents/NewsReleases/2003/09/0328.doc; News Release, *Veneman Announces $8.8 Million in Broadband Grants*, USDA, Release No. 0480.04 (Oct. 29, 2004), http://www.usda.gov/wps/portal/!ut/p/s.70A/701RD?printable=true&contentidonly=true&contentid=2004/10/0480.xml; News Release, *USDA Announces $9 Million in Broadband Grants*, USDA, Release No. 0409.05 (Sept. 30, 2005), http://www.usda.gov/wps/portal/!ut/p/s.70A/701RD?printable=true&contentidonly=true&contentid=2005/09/0409.xml; News Release, *USDA Announces $8.9 Million in Broadband Community Connect Grants*, USDA, Release No. 0360.06 (Sept. 15, 2006), http://www.usda.gov/wps/portal/!ut/p/s.70A/701RD?printable=true&contentidonly=true&contentid=2006/09/0360.xml; News Release, *USDA Announces Recipients of Community Connect Grants*, USDA, Release No. 0279.07 (Oct. 4, 2007), http://www.usda.gov/wps/portal/!ut/p/s.70A/701RD?printable=true&contentidonly=true&contentid=2007/10/0279.xml.

57 News Release, *USDA Announces Recipients of Community Connect Grants*, USDA, Release No. 0279.07 (Oct. 4, 2007), http://www.usda.gov/wps/portal/!ut/p/s.70A/701RD?printable=true&contentidonly=true&contentid=2007/10/0279.xml.

58 *Id.*

59 News Release, *USDA Awards More Than $22.3 Million for Distance Learning and Telemedicine Grants*, USDA, Release No. 0275.07 (Oct. 1, 2007), http://www.usda.gov/wps/portal/!ut/p/s.70A/701RD?printable=true&contentidonly=true&contentid=2007/10/0275.xml.

60 On this point, the FCC has recently proposed steps to improve the quality and depth of its data collection concerning broadband deployment. *See Development of Nationwide Broadband Data to Evaluate Reasonable and Timely Deployment of Advanced Services to All Americans, Improvement of Wireless Broadband Subscribership Data, and Development of Data on Interconnected Voice over Internet Protocol (VoIP) Subscribership*, WC Doc. No. 07-38, Notice of Proposed Rulemaking, 22 FCC Rcd 7760 (2007). The more complete and granular data sought by the Commission should yield a much higher resolution picture of the broadband landscape that will benefit all parties interested in broadband deployment.

61 Specifically, the FCC reports that there were 82,547,651 high-speed lines in the United States in December 2006, up from 6,756,877 in December 2000. *See High-Speed Services for Internet Access: Status as of December 31, 2006*, Federal Communications Commission (Oct. 2007), Table 1 (FCC Study).

62 *Id.* at Table 3, Chart 5.

63 *Internet Use Supplement to the October 2007 CPS*, U.S. Census Bureau (October 2007 CPS). This represents the seventh such study by the Commerce Department involving data collection by the Census Bureau, and analysis by NTIA, and (on three occasions) its sister agency, the Economics and Statistics Administration (ESA). *See* http://www.ntia.doc.gov/reports/anol/index.html.

64 The broadband data collected by the U.S. Census Bureau in its *Internet and Computer Use Supplement to the September 2001 CPS*, and relied on herein, is set forth in four tables developed by the Census Bureau, excerpts of which appear at Appendix A of this report. Table A-1 provides data for total U.S. households for Internet connection type. Table A-2 provides data for total rural households by Internet connection type. Table A-3 provides data for total U.S. households with Internet. Table A-4 provides data for rural households with Internet. The referenced tables in their entirety are available on NTIA's website at http://www.ntia.doc.gov/reports/2008/NetworkedNation.html. The figure of 9.1 percent represents the sum of households with DSL (3,335,000) and cable modem services (6,594,000) from Table A-1 expressed as a percentage of the 108,562,000 total Internet households in the United States from Table A-3. Note: Table A-1 also includes a very small

"Other" category (representing .85 percent of all U.S. Internet households). The data does not clearly identify whether these "Other" types of Internet connections (i.e., non dial- up, DSL, or cable modem) were broadband in nature. Accordingly, this category was excluded from the 2001 broadband percentage computation.

65 The broadband data collected by the Census Bureau from the October 2007 CPS, and relied on herein, is set forth in three tables developed by the Census Bureau, excerpts of which appear at Appendix B of this report. Table B-1 provides data for total U.S. households using the Internet. Table B-2 provides data for total rural households using the Internet. Table B-3 provides data for U.S. households with Internet broadband access ranked by States. The tables from which the broadband data was excerpted are printed in their entirety on NTIA's website at http://www.ntia.doc.gov/reports/2008/NetworkedNation.html. The figure of 50.8 percent appears in Table B-1. In the October 2007 CPS, 59,847,000 households (out of a total U.S. figure of 117,840,000 responding to the survey) reported having a "broadband" connection of some kind. Respondents using the Internet at home were asked whether they had (1) a "dial-up" connection; (2) a "broadband" connection; or a (3) "something else" (i.e., "Other").

66 *See* Appendix A, Table A-2, Table A-4. The figure of 5.6 percent represents the sum of rural households with DSL (339,000) and cable modems (1,151,000) from Table A-2 as a percentage of the 26,832,000 rural households in the United States from Table A-4. Again, as noted *supra* note 65, data for respondents in the "Other" category (.75 percent of rural Internet households) was excluded due to uncertainties relative to the broadband status of these households.

67 *See* Appendix B, Table B-2. In the October 2007 CPS, 9,072,000 rural households (out of a total of 23,394,000 rural households covered by the survey) reported having some type of broadband connection at home.

68 For example, the percent of Internet households with dial-up connectivity declined from 81.0 percent nationwide in September 2001 to 17.3 percent as of October 2007 (equals 12,575,000 dial-up households out of a total of 72,721,000 Internet households in the United States). *See* Appendix A, Table A-1; Appendix B, Table B-1. In rural areas, the dial-up proportion declined from 87.8 percent to 33.1 percent (equals 4,514,000 dial-up households out of 13,646,000 total rural Internet households) during the same period. *See* Appendix A, Table A-2; Appendix B, Table B-2.

69 *See* Appendix B, Table B-3. This figure is calculated from October 2007 CPS data (based on 90 percent confidence intervals used in sampling, *i.e.*, the 90

percent probability that the actual number for a given state in this case lies within a calculated range).

70 *Id.* Specifically, the percentages derived from the CPS sample are 64.9 percent for New Hampshire and 62.5 percent for Alaska. Statistically, there is a 90 percent confidence that the actual percentage lies in the range 62.1 to 67.7 percent for New Hampshire and 59.6 to 65.5 percent for Alaska.

71 FCC Study, *supra* note 62, Table 3.

72 Comments of Verizon and Verizon Wireless in GN Docket No. 07-45, Inquiry Concerning the Deployment of Advanced Telecommunications Capability to All Americans in a Reasonable and Timely Fashion, and Possible Steps to Accelerate Such Deployment Pursuant to Section 706 of the Telecommunications Act of 1996 (May 16, 2007) at 11 & n. 32 (Verizon/VZW Comments), http://fjallfoss.fcc.gov/prod/ecfs/retrieve.cgi?nativeorpdf=pdf&iddocument=6519411810 (*citing* J. Hodulik, et al., UBS, Qwest Communications at Table 1 (Oct. 4, 2005) (weighted average).

73 Comments of the Organization for the Promotion and Advancement of Small Telecommunications Companies in GN Docket No. 07-45, Inquiry Concerning the Deployment of Advanced Telecommunications Capability to All Americans in a Reasonable and Timely Fashion, and Possible Steps to Accelerate Such Deployment Pursuant to Section 706 of the Telecommunications Act of 1996 (May 16, 2007) at 3-4 (OPASTCO Comments), http://fjallfoss.fcc.gov/prod/ecfs/retrieve.cgi?nativeorpdf=pdf&iddocument=6519411737. OPASTCO surveyed 310 of its member companies of which 72 responded. *Id.* at 3.

74 National Exchange Carrier Association, *DSL Data Collection Summary*, http://www.neca.org/source/NECAPublications4943.asp (last visited Jan. 15, 2008).

75 *Id.*

76 *NTCA 2006 Broadband/Internet Availability Survey Report,* National Telecommunications Cooperative Association (Aug. 2006) at 3, http://fjallfoss.fcc.gov/prod/ecfs/retrieve.cgi?nativeorpdf=pdf&iddocument=6519411694 (last visited Jan. 15, 2008). NTCA surveyed about 570 members companies of which 120 responded. *Id.*

77 Verizon/VZW Comments, *supra* note 722, at 14.

78 Comments of AT&T in GN Docket No. 07-45, Inquiry Concerning the Deployment of Advanced Telecommunications Capability to All Americans in a Reasonable and Timely Fashion, and Possible Steps to Accelerate Such

Deployment Pursuant to Section 706 of the Telecommunications Act of 1996 (May 16, 2007) at 7 (AT&T Comments), http://fjallfoss.fcc.gov/prod/ecfs/retrieve.cgi?nativeorpdf=pdf&iddocument=6519411743.

79 OPASTCO Comments, *supra* note 73, at 3.

80 *Broadband Connectivity Competition Policy*, Federal Trade Commission at 101 (June 2007) (FTC Report), http://www.ftc.gov/reports/broadband/v070000report.pdf (*citing* Bernstein Research, Broadband Update: "Value Share" and "Subscriber Share" Have Diverged at 4 (2006)). Please note that the sources from which the price data is cited throughout this Report do not indicate that the data has been adjusted for inflation.

81 AT&T Comments, *supra* note 788, at 7. *See also* Verizon/VZW Comments, *supra* note 722, at 13-14 (ILEC's average DSL prices have fallen approximately 30 percent over the past four years).

82 FTC Report, *supra* note 81, at 101.

83 FCC Study, *supra* note 61, Table 3. "High-speed" lines must have transmission speed in excess of 200 Kbps in at least one direction.

84 Chart 2 compiles ADSL growth data that appear in tabular form in FCC Study, Tables 1-4. It is important to note that "advanced" services lines represent a subset of total high-speed lines. The data presented in Chart 2 were derived as follows: *(1)* the total number of "advanced services lines" [FCC Table 2] were subtracted from the total number of all high-speed lines [from FCC Table 1] to yield the total net number of "basic" high-speed lines; *(2)* "residential advanced services" lines [FCC Table 4] were then subtracted from total "residential high-speed lines" [FCC Table 3] to yield the total net number of *residential "basic" high-speed lines*; *(3)* the result of step two (net residential basic) was then subtracted from the result of step one (total net basic) to determine the total net number of *non-residential (i.e., business) "basic" high-speed lines. (4)* Next, "residential advanced services" lines [FCC Table 4] were subtracted from total "advanced service lines" [FCC Table 2] to determine the total net number of *non-residential (i.e., business) "advanced services lines."* The 22.8 million residential high-speed lines cited in the text represent the sum of residential "basic" high-speed lines and residential "advanced services lines," which are depicted in the dark blue and light blue areas of the chart. The layering of the chart reveals not only the significant growth that has occurred in high-speed lines overall, but also the even greater relative increase in the number of "advanced services lines" as proportion of the total.

85 *Id.*, Table 1.

86 *Id.*, Table 4. Again, asymmetric services represent virtually all advanced DSL lines. Advanced services must offer transmission speeds in excess of 200 Kbps in both directions.

87 *Id.*, Table 2.

88 See discussion of targeted deregulation initiatives and elimination of artificial distinctions, *supra* pp. 7-8.

89 T. Hazlett, *Empirical Evidence on the Effect of Broadband Regulation*, Powerpoint presentation at George Mason University Law School Information Economy Project Mini Conference: "Lessons from the Telecom Wars" (Sept. 28, 2006) at 15, http://mason.gmu.edu/~thazlett/Conferences/LessonsfromtheTelecomWarsPresentations.html (last visited Jan. 15, 2008). Professor Hazlett attributes the increase in DSL growth to the FCC's elimination of "line sharing." *See* Verizon/VZW Comments, *supra* note 722, at 6.

90 J. Eisenach, Ph.D., Testimony before the Senate Committee on Commerce, Science, and Transportation at 2 and Figure 1 (April 24, 2007), http://commerce.senate.gov/public/files/TestimonyJeffreyEisenachCriterionEisenachTestimonyWithFigures042407.pdf.

91 *See* NCTA Website, Broadband Internet Access, http://www.ncta.com/IssueBrief.aspx?contentID=3024 (last visited Jan. 24, 2008).

92 NCTA Comments, *supra* note 91, at 8.

93 *Id.*

94 *See* TIA Market Study, *supra* note 18, at 56-57.

95 Cable, *Communications Daily* (May 23, 2007), at 15.

96 TIA Market Study, *supra* note 18, at 5.

97 *Id.* at 4, 6.

98 *Id.* at 12, 13 ("The advent of new applications is driving demand for new devices. Subscribers must upgrade their devices to use new services such as picture messaging, mobile music, 3-D video games, music videos, television, children's location services, instant messaging, and Internet access.").

99 FCC Study, *supra* note 61, Table 1.

100 *Id.*, Table 2.

101 *Annual Report and Analysis of Competitive Market Conditions With Respect to Commercial Mobile Services,* Implementation of Section 6002(b) of the Omnibus Budget Reconciliation Act of 1993, WT Doc. No. 06-17, 21 FCC Rcd 10947, 10950 ¶ 3 (2006) (FCC 11t[h] Annual CMRS Competition Report).

102 *Id.* (emphasis added, internal footnote omitted).

103 News Release, *AT&T, Pantech Double-Up on Cool with Slim, Dual-Keyboard Pantech Duo, New 3G Windows Mobile 6 Global Smart Device*

Bridges Mobile Needs of Personal and Professional Lifestyles With Variety of Entertainment and Business Features (Oct. 29, 2007), http://www.att.com/gen/press-room?pid=4800&cdvn=news&newsarticleid=24621 (last visited Jan. 15, 2008).

104 News Release, *Verizon Wireless: 100 Percent of Wireless Broadband Network Now Enhanced with Faster Speeds,* Verizon Wireless (June 29, 2007), http://news.vzw.com/news/2007/06/pr2007-06-28h.html (last visited Jan. 15, 2008).

105 News Release, *Central Valley Can Do More with Wireless: Faster Sprint Mobile Broadband Expands in California,* Sprint (Oct. 1, 2007), http://newsreleases.sprint.com/phoenix.zhtml?c=127149&p=irol-newsArticlePrint&ID=1057229&highlight (last visited Jan. 15, 2008).

106 Comments of Sprint Nextel Corporation in FCC GN Docket No. 07-45, Inquiry Concerning the Deployment of Advanced Telecommunications Capability to All Americans in a Reasonable and Timely Fashion, and Possible Steps to Accelerate Such Deployment Pursuant to Section 706 of the Telecommunications Act of 1996 (May 16, 2007), at 7 (Sprint Nextel Comments).

107 T. von Fuchs, *T-Mobile Denies 3G Rumor*, Wireless Week (Nov. 28, 2007), http://www.wirelessweek.com/T-Mobile-3G-Rumor.aspx (last visited Jan. 15, 2008).

108 News Release, *Deutsche Telekom Well Positioned for Further Growth in the United States*, 3G Americas (Oct. 6, 2006), http://www.3gamericas.org/English/Newsroom/DisplayPressRelease.cfm?id=2820&s=ENG (last visited Jan. 15, 2008).

109 *See* News Release, *Unlicensed Mobile Access System Solution from NXP Supports First Nationwide U.S. Commercial Rollout of FixedMobile Convergence*, NXP (July 11, 2007), http://www.businesswire.com/portal/site/google/index.jsp?ndmViewId=newsview&newsId=20070711005350&newsLang=en (NXP News Release) (last visited Jan. 15, 2008).

110 FCC 11th Annual CMRS Competition Report, 21 FCC Rcd at 11028-29, ¶ 210.

111 TIA Market Study, *supra* note 18, at 69.

112 FCC Study, *supra* note 61, Table 1.

113 TIA Market Study, *supra* note 18, at 70.

114 *Id.*

115 *Id.* at 69, 186.

116 FCC 11th Annual CMRS Competition Report, 21 FCC Rcd at 11028-29, ¶ 210 (footnotes omitted).

117 TIA Market Study, *supra* note 18, at 194.

118 *Id.* at 195.

119 *Id.*

120 *Id.* at 196.

121 T-Mobile USA, Customer Connection Center, US Locations, https://selfcare.hotspot.tmobile.com/locations/viewLocationMap.do (last visited Dec. 27, 2007).

122 UMA Today, UMA Technology: Extending Mobile Services to Unlicensed Spectrum; UMA Overview, http://www.umatechnology.org/overview/ (last visited Dec. 27, 2007).

123 *Id.*

124 NXP News Release, *supra* note 110.

125 T-Mobile HotSpot @Home – General Information, http://support.t-mobile.com/knowbase/root/public/tm23449.htm?&printpage=1 (last visited Jan. 15, 2008).

126 *Id.; see also* J. Cheng, *T-Mobile First in US with Long-awaited UMA cell-to-WiFi Service,* Ars Technica (June 27, 2007), http://arstechnica.com/news.ars/post/20070627-t-mobile-intros-long-awaited-cell-to-wifi-service.html (last visited Jan. 15, 2008).

127 E. Malykhina, *Qualcomm, Nortel Team on Dual-Mode Phone Service*, Information Week (December 3, 2007), http://www.informationweek.com/internet/showArticle.jhtml?articleID=204600791&subSection=Internet (last visited Jan. 15, 2008).

128 FTC Report, *supra* note 80, at 106.

129 TIA Market Study, *supra* note 18, at 69, 197.

130 News Release, *Municipal Wi-Fi Needs a Paradigm Shift, Says ABI Research*, Business Wire (Dec. 4, 2007), http://www.businesswire.com/portal/site/google/index.jsp?ndmViewId=newsview&newsId=20071204005788&newsLang=en (last visited Jan. 15, 2008).

131 FTC Report, *supra* note 80, at 106.

132 R. Martin, *EarthLink's Cutbacks Cast a Cloud Over Muni Wi-Fi*, Information Week (Sept. 1, 2007), http://www.informationweek.com/news/showArticle.jhtml?articleID=201803540 (last visited Jan. 15, 2008).

133 *Id.*; *see also* TIA Market Study, *supra* note 18, at 197-98 (noting projects in Anaheim, California; Corpus Christi, Texas; Annapolis, Maryland, St. Cloud, Florida, and Tempe, Arizona).

134 I. Groves, *Cities Want to Bring Wireless Internet to Downtowns*, Burlington Times-News (Nov. 12, 2007), http://www.thetimesnews.com/articles/city7521article.html/wifiinternet.html (last visited Jan. 15, 2008).

135 M. St. Aude, *Wi-Fi 'Cloud' to Blanket County by 2009*, Casa Grande Valley Newspapers (Nov. 28, 2007), http://www.zwire.com/site/news.cfm?newsid=19066431&BRD=1817&PAG=461&deptid=222089&rfi=6 (last visited Jan. 15, 2008).

136 News Release, *Community Connect Awards – Mississippi*, USDA Rural Development, http://www.usda.gov/rus/telecom/commconnect/pdfs/ms-commconnectsummaries.pdf. A cotton gin building owned by the city will be used to set up a community center, which will be used to offer a variety of educational and training services to residents. Residents will benefit from better educational and economic opportunities by improved access to information, prescription medicines, job search and GED college correspondence courses. Businesses will benefit from fast and reliable access to information, which can increase their productivity.

137 R. Martin, *EarthLink's Cutbacks Cast a Cloud Over Muni Wi-Fi*, *supra* note 133.

138 J. Merron, *Municipal Wi-Fi: A Promise Unfulfilled?*, Information Week (Nov. 6, 2007), http://www.informationweek.com/mobility/showArticle.jhtml?articleID=202802427&subSection=Wi-Fi/WiMax (last visited Jan. 15, 2008); *see also* D. Jones, *Moto Muni: So Much for the City*, Unstrung (Nov. 28, 2007), http://www.unstrung.com/document.asp?docid=139892&print=true (last visited Jan. 15, 2008).

139 TIA Market Study, *supra* note 18, at 156, 162.

140 *Id.* at 186, 187.

141 *Id.* at 187.

142 News Release, *AT&T Initiatives Expand Availability of Advanced Communications Technologies,* AT&T (May 8, 2006), http://www.att.com/gen/press-room?pid=5097&cdvn=news&newsarticleid=22272 (last visited Jan. 15, 2008); and *AT&T WiMAX and Fixed Wireless Deployment Fact Sheet*, http://www.att.com/Common/files/pdf/WiMAXFactSheet.pdf (last visited Jan. 15, 2008).

143 News Release, *ITU-R Radiocommunication Assembly Expands IMT-2000 3G Radio Interface Family with OFDMA technology; Establishes IMT-Advanced as the name for 4G*, ITU (Oct. 19, 2007), http://www.itu.int/newsroom/wrc/2007/itur web flash/20071019.html (last visited Jan. 15, 2008). *See also* Associated Press, *UN Meeting Gives Mobile Service Providers Access to Terrestrial TV Spectrum* (Nov. 15, 2007), http://nl.newsbank.com/nl-search/we/Archives?paction=list&ptopdoc=11 (last visited Jan. 15, 2008) (password protected); Telecommunications Reports, *Identification of IMT Spectrum, 2.5 GHz Band Protection Seen As*

Crucial Decisions Made at WRC-07 in Geneva, vol. 73, no. 23 at 3 (Dec. 1, 2007).

144 News Release, *Clearwire Brings High-Speed Wireless Broadband and Phone Service to "The Queen City,"* Clearwire Corp. (Dec. 3, 2007), http://newsroom.clearwire.com/phoenix.zhtml?c=214419&p=irol-newsArticle&ID=1083314 (last visited Jan. 15, 2008).

145 Clearwire Corp. Coverage Area and Maps, http://www.clearwire.com/store/serviceareas.php (last visited Jan. 15, 2008).

146 *See* Comments of Clearwire Corp. in FCC GN Docket No. 07-45, Inquiry Concerning the Deployment of Advanced Telecommunications Capability to All Americans in a Reasonable and Timely Fashion, and Possible Steps to Accelerate Such Deployment Pursuant to Section 706 of the Telecommunications Act of 1996 (May 16, 2007) at 6 (Clearwire Comments).

147 *Id.* at 2, 4. As of May 2007, Clearwire was also offering portable VoIP service over its network in roughly 26 of its markets, with plans to expand to others. *Id.* at 2, n. 4.

148 *Id.* at 4, 5.

149 B. Gardiner, *Intel Says WiMax Is Coming Next Year – This Time for Sure*, Wired (Sept. 21, 2007), http://www.wired.com/techbiz/it/news/2007/09/intelwimax (last visited Jan. 15, 2008).

150 *Id.*

151 *Id.*

152 TIA Market Study, *supra* note 18, at 187.

153 W. D. Gardner and E. Malykhina, *WiMax Gets Two New Friends: Cisco and the United Nations*, Information Week (Oct. 27, 2007), http://www.informationweek.com/mobility/showArticle.jhtml?articleID=202602151&subSection=Wi-Fi/WiMax (last visited Jan. 15, 2008).

154 News Release, *Sprint Nextel Announces 4G Wireless Broadband Initiative with Intel, Motorola and Samsung*, Sprint Nextel (August 8, 2006), available at: http://www2.sprint.com/mr/newsdtl.do?id=12960 (last visited Jan. 15, 2008).

155 *Id.*

156 J. Gertzen, *Sprint Seeks Outside Funding for WiMax Plans*, Kansas City Star (Dec. 5, 2007), at C1.

157 *Id.*; *see also* M. Hamblen, *Report: Sprint Exec Says a Wimax Service Could be Spun Off; It's Looking to Save Money Needed for a New Network*, Computerworld (Dec. 4, 2007), http://www.computerworld.com/action/article.do?command=viewArticleBasic&articleId=9050961&intsrc=hmlist (last visited Jan. 15, 2008).

158 *See* FCC Study, *supra* note 61, Table 16.

159 *See generally,* Hughes Communications, Inc., 2006 Annual Report 5 (2007) http://www.hughes.com/HUGHES/Doc/0/DAM4EI2JNF14D2TGHII4VU9F7E/HCI2006AnnualReport.pdf (Hughes Annual Report); and WildBlue Communications, Inc., Questions & Answers, http://wildblue.com/aboutWildblue/qaa.jsp#11 (last visited Jan. 15, 2008); and WildBlue Communications, Inc., How it Works Demo, http://wildblue.com/aboutWildblue/howitworksdemo.jsp (last visited Jan. 15, 2008).

160 BGAN Global Voice and Broadband Data, Overview Brochure, Inmarsat, http://www.inmarsat.com/Downloads/English/BGAN/Collateral/bganoverviewbrochureEN.pdf?language=EN&textonly=Fals e, at 6 (last visited Jan. 15, 2008).

161 TIA Market Study, *supra* note 18, at 71 and Table II-2.36. *Cf.* Comments of The Satellite Industry Association (SIA) in WC Docket No. 07-52, *Broadband Industry Practices* June 15, 2007), at 2 (places the number of satellite broadband consumers at over 500,000) (SIA Comments), http://fjallfoss.fcc.gov/prod/ecfs/retrieve.cgi?nativeorpdf=pdf&iddocument=6519532283.

162 W. J. Holstein, *Sending Up Satellites and Closing the Digital Divide*, New York Times (Oct. 20, 2007) (interview with Pradman P. Kaul, chief executive of Hughes Communications, Inc.), http://www.nytimes.com/2007/10/20/technology/20interview.html (last visited Jan. 15, 2008).

163 Hughes Annual Report, supra note 159, at 5. *See also* Annual Report and Analysis of Competitive Market Conditions with Respect to Domestic and International Satellite Communications Services, IB Doc. No. 06-67, First Report, 22 FCC Rcd 5954, 6011 ¶ 186 (2006).

164 *See* SIA Comments, *supra* note 161, at 2 ("The satellite industry currently provides competitive broadband services throughout the United States in the 48 contiguous states, the District of Columbia, Alaska, Hawaii, Puerto Rico and the US Virgin Islands."). Hughes reports that "HughesNet service coverage is continent-wide, reaching an estimated 10 million or more households and over 3 million small businesses in the U.S. that are unserved by terrestrial broadband technologies." Hughes Annual Report, *supra* note 159, at 5. *See also,* W. J. Holstein, *Sending Up Satellites and Closing the Digital Divide*, New York Times (Oct. 20, 2007), http://www.nytimes.com/2007/10/20/technology/20interview.html (last visited Jan. 15, 2008).

165 J. Barthold, *TerreStar's Goal: Define, Create 4G Network*, Telecommunications Online (Sept.19, 2007), http://www.telecommagazine.

com/article.asp?HHID=AR3484 (last visited Dec. 27, 2007); *TerreStar-1 to Launch in Late 2008*, Sat- News Daily (May 31, 2007), http://www.satnews.com/stories2007/4529/ (last visited Jan. 15, 2008); *TerreStar Granted Spectrum for Hybrid Satellite-Terrestrial Network*, SatNews Daily (May 21, 2007), http://www.satnews.com/stories2007/4481/ (last visited Jan. 15, 2008).

166 News Release, *Launch Dates Agreed with ILS for Third Inmarsat-4 Satellite*, Inmarsat (July 8, 2007), http://www.inmarsat.com/About/Newsroom/00023055.aspx?language=EN&textonly=False (last visited Jan. 15, 2008); *Our Satellites: Most Advanced of Their Kind*, Inmarsat, http://www.inmarsat.com/About/Oursatellites/default.aspx?language=EN&textonly=False (last visited Jan. 15, 2008).

167 News Release, *MSV, ILS Contract to Launch Next-Generation Satellite,* Mobile Satellite Ventures (May 15, 2007), http://www.msvlp.com/media/press-releases-view.cfm?id=126&yr=2007 (last visited Jan. 15, 2008).

168 *See Next Generation Network, Differentiation from Existing MSS*, Mobile Satellite Ventures, http://www.msvlp.com/nextgen/differentiation.cfm (last visited Jan. 15, 2008) (discussing MSV's objective of achieving transparency in its next-generation integrated satellite/ATC network).

169 *Broadband Deployment is Extensive Throughout the United States, But It Is Difficult to Assess the Extent of Deployment Gaps in Rural Areas*, Government Accountability Office, GAO-06-426 (May 2006) at 23, http://www.gao.gov/new.items/d06426.pdf (last visited Jan. 15, 2008).

170 FCC Study, *supra* note 61, Table 1.

171 *Id.* at Table 1 and accompanying note 2.

172 TIA Market Study, *supra* note 18, at 84-85.

173 *Id.* at 84.

174 *Id.* at 85.

175 *Id.* at 84.

176 *Id.* at 43, 61, 84.

177 *Id.* at 84.

178 *Availability of Advanced Telecommunications Capability in the United States; Fourth Report to Congress*, Federal Communications Commission (Sept. 9, 2004) at 16, http://hraunfoss.fcc.gov/edocspublic/attachmatch/FCC-04-208A1.pdf.

179 *Id.*

180 News Release, *Fiber to the Home Revs Up Expansion, More than Two Million Homes Now Connected to Next-Generation Broadband*, Fiber-to-the-

Home Council (Oct. 3, 2007), http://www.ftthcouncil.org/?t=277 (last visited Jan. 15, 2008).

181 *Id.*

182 *Id.*

183 *Id.*

184 *Id.*

185 News Release, *Verizon Continues to Dramatically Raise Broadband Upload Speeds in FiOS Internet Service Areas,* Verizon (Nov. 20, 2007), http://newscenter.verizon.com/press-releases/verizon/2007/verizon-continues-to.html (last visited Jan. 15, 2008).

186 *Id.*

187 *Id. See also* Verizon/VZW Comments, *supra* note 722, at 1, 7.

188 TIA Market Study, *supra* note 18, at 61; *see also* Fact Sheet, *How AT&T U-verse TV is Delivered*, AT&T, http://www.att.com/Uverse/files/How UverseIsDelivered2-22.pdf (last visited Jan. 15, 2008).

189 Fact Sheet, *How AT&T U-verse TV is Delivered*, AT&T, http://www.att.com/Uverse/files/HowUverseIsDelivered2-22.pdf (last visited Jan. 15, 2008).

190 Media Kits, *AT&T U-verse, Innovative New Entertainment Experience*, AT&T, http://www.att.com/gen/press-room?pid=5838 (last visited Jan. 15, 2008).

191 *Id.*

192 *Id.*

193 *AT&T Raises Spending Outlook for U-verse Expansion*, Reuters (Nov. 6, 2007), http://www.reuters.com/article/technologymedia-telco-SP/idUSN0 640400220071 106 (last visited Jan. 15, 2008).

194 TIA Market Study, *supra* note 18, at 84-85.

195 *Id.* at 59, 84.

196 *Id.* at 61, 84.

197 *Id.* at 84-85.

198 *Id.* at 84.

199 MV lines carry electricity from a local substation, where power from regional and interstate power lines is reduced to a voltage that can reliably and safely be transmitted to residential and business customers. In the United States, a substation will typically serve about 50 homes and MV power lines run between 9 and 30 miles. The LV lines are roughly equivalent to a telephone companies' "loop" facilities, and carry power from the last utility pole to a customer's home or business. At some point, the BPL provider extracts broadband signals from the power lines and, after aggregating traffic from

multiple homes, processing the data, and adding routing information, connects them to a communications "backhaul" network (*e.g.*, the Internet or the public telephone network) using wired or wireless facilities. For a more complete description of BPL technology, *see* L. Lushbaugh and S. Safavian, *Broadband Over Power Lines (BPL)*, 5 Bechtel Telecommunications Technical Journal 19, 22-26 (Jan. 2007).

200 *Status of Broadband Over Powerline 2007*, United Power Line Council, http://uplc.utc.readyportal.net/file depot/0-10000000/0-10000/7966/conman/ 2007+BPL+Update.pdf (last visited Jan. 15, 2008).

201 E. Gubbins, *Powerline Promises Broken on Broadband*, Telephony Online (May 22, 2006), http://telephonyonline.com/mag/telecompowerline promisesbroken/ (last visited Jan. 15, 2008).

202 *TXU Electric Delivery Strives to Improve Reliability with BPL*, Transmission & Distribution World (April 1, 2007), http://tdworld. com/customerservice/powertxuelectricdelivery-bpl/ (last visited Jan. 15, 2008). Please note that Duke Energy acquired Cinergy in 2006.

203 K. Belson, *Power Companies Enter the High-Speed Market*, New York Times (Oct. 17, 2005), http://www.intellon.com/pdfs/bplNYTimes 20051015.pdf (last visited Jan. 15, 2008).

204 TIA Market Study, *supra* note 18, at 69; *TXU Electric Delivery Strives to Improve Reliability with BPL*, *supra* note 202.

205 FCC Study, *supra* note 61, Table 6.

206 TIA Market Study, *supra* note 18, at 69.

207 C. Roden, *FTTx and BPL: Analysis and Outlook, First Quarter 2007*, Parks Associates, http://www.parksassociates.com/research/reports/tocs/pdfs/parks-FTTXandBPL.pdf (last visited Jan. 15, 2008).

208 E. Gubbins, *supra* note 201 (*quoting* Nicole Klein of the Yankee Group).

209 *Id.* (*citing* ComTek Chief Executive Officer Joe Fergus).

210 Posting of M. Arden to Analyst Industry Blogs, *Earthlink's Broadband Strategy*, ABI Research, http://www.abiresearch.com/Blog/DigitalHome Blog/127 (Oct. 18, 2006, 3:47 PM) (last visited Jan. 15, 2008).

211 E. Bangeman, *DIRECTV Plugs into Broadband Over Power Line*, Ars Technica (Aug. 15, 2007), http://arstechnica.com/news.ars/post/20070815-directv-plugs-into-broadband-over-power-line.html (last visited Jan. 15, 2008).

212 FCC Study, *supra* note 61, at 4 and Table 16.

213 Verizon/VZW Comments, *supra* note 722, at 6.

214 *National Cable & Telecommunications Association* (Kagan Research estimate), http://www.ncta.com/ContentView.aspx?contentId=56 (last visited Jan. 15, 2008).

215 NCTA Comments, *supra* note 91, at 6-7.

216 Cable, *Communications Daily* at 15 (May 23, 2007).

217 TIA Market Study, *supra* note 18, at 84.

218 *Id.*

219 *Id.* at 162.

220 *Id.*

221 M. Bruun, *Wireless is More,* Worcester (MA) Telegram & Gazette (Nov. 6, 2007), http://www.telegram.com/article/20071106/NEWS/711060630/1008/NEWS02 (last visited Jan. 15, 2008).

222 TIA Market Study, *supra* note 18, at 162.

223 Verizon/VZW Comments, *supra* note 72, at 7.

224 *AT&T Raises Spending Outlook for U-verse Expansion*, Reuters (Nov. 6, 2007), http://www.reuters.com/article/technology-media-telco-SP/idUSN0640400220071106 (last visited Jan. 15, 2008); *see also* Reply Comments of AT&T, Inc. in GN Docket No. 07- 45, Inquiry Concerning the Deployment of Advanced Telecommunications Capability to All Americans in a Reasonable and Timely Fashion, and Possible Steps to Accelerate Such Deployment Pursuant to Section 706 of the Telecommunications Act of 1996 (May 31, 2007) at 4 (AT&T Reply Comments).

225 J. Bartash, *Qwest's Earning Soar on Tax Gain*, MarketWatch (Oct. 30, 2007), http://www.marketwatch.com/News/Story/qwests-earnings-soar-tax-gain/story.aspx?guid=%7BA0FAD201-E0ED-46FF-AB02-82051EAF65F4%7D (last visited Jan. 15, 2008).

226 FTC Report, *supra* note 81, at 101.

227 *Id.* at n. 469 (*quoting* Robert W. Hahn & Robert E. Litan, *The Myth of Network Neutrality and What We Should Do About It* 10 (AEI-Brookings Joint Center, Working Paper No. RP-06-33, Nov. 2006)), http://www.aei-brookings.org/admin/authorpdfs/page.php?id=1357 ("While the absolute price of a cable modem has not declined as rapidly, the quality-adjusted price has declined significantly, as cable modem connection speeds have more than doubled while prices held steady.").

228 TIA Market Study, *supra* note 18, at 67.

229 Verizon/VZW Comments, *supra* note 722, at 7 & n. 22 (*citing* D. Barden, *et al.*, Bank of America, *Battle for the Bundle: Consumer Wireline Services Pricing* at 9 (Jan. 23, 2006) ("The rollout of Verizon's FiOS service in select

markets has elicited thinly advertised, yet highly competitive pricing responses for incumbent cable providers... . In each of these markets the respective cable provider ... has responded with competitive pricing, well below their national average."))

230 Comments of the National Rural Telecommunications Cooperative in USDA Rural Utilities Service Docket No. RUS-06-Agency-0052 (July 10, 2007) at 5, http://www.nrtc.coop/export/main/newspolicy/NRTCCommentsRUS broadband071007.pdf (NRTC RUS Comments).

231 *Wildblue Slashes Installation Price, Adds $200 Million to Coffers*, Satellite News (Jan. 23, 2006), http://www.accessmylibrary.com/coms2/summary 0286-12484725ITM (last visited Jan. 15, 2008).

232 SIA Comments, *supra* note 161, at 3 n. 4.

233 *Id.* Satellite broadband market analysts have echoed this analysis. *See Wildblue Slashes Installation Price, Adds $200 Million to Coffers*, *supra* note 231 ("'I expect that Wildblue would like to get the installation price down to the point where they could cover it entirely, the way that DirecTV does,' [Frost & Sullivan analyst, Max] Engel said. To do this, Wildblue will need to sell enough home satellite terminals to drive down their manufacturing costs substantially. 'It's a matter of the more people they get, the cheaper they can offer the product, and the cheaper they can offer the product, the more people they can get.'").

234 *See* NRTC RUS Comments, *supra* note 230, at 2 & n. 2. *See also* Wildblue Slashes Installation Price, Adds $200 Million to Coffers, *supra* note 231 (also observing that "Wildblue has signed up five more value-added resellers to help move its enterprise VSAT products ... [which are] aimed at telecommuters, retailers, financial services companies, government, telemedicine, and disaster recovery providers, among others."); and *While GAO Fiddles, Wildblue, Hughes Burn Ahead on Broadband*, Satellite News (May 22, 2006),http://www.accessmylibrary.com/coms2/summary0286-15329851ITM (last visited Jan. 15, 2008).

235 O. Kharif, *Wal-Mart's Latest Sale: Broadband*, BusinessWeek (Oct. 8, 2007), http://www.businessweek.com/technology/content/oct2007/tc20071080600026 .htm (last visited Jan. 15, 2008); *see also Broadband Across America: Through Wal-Mart, Hughes Brings High-Speed Internet to Rural Communities*, Hughes Communications, Inc. (Oct. 9, 2007), http://www.hughes.com/HUGHES/Doc/0/PMG7IF1L6RL456TDI3HPA5SK 14/100907.htm. (last visited Jan. 15, 2008).

236 O. Kharif, *Wal-Mart's Latest Sale: Broadband*, BusinessWeek, *supra* note 235.

[237] *See* HughesNet, "Pricing," http://go.gethughesnet.com/HUGHES/Rooms/DisplayPages/LayoutInitial?Container=com.webridge.entity.Entity%5BOID%5B71 A9F5B422ABCE4886D9492F66B5B589%5D%5D (last visited Jan. 15, 2008).

[238] *Id*.

[239] *See* FCC Study, *supra* note 61, at 4 & Table 16.

APPENDIX A: INTERNET USE SUPPLEMENT TO THE SEPTEMBER 2001 CURRENT POPULATION SURVEY. U.S. CENSUS BUREAU (SEPTEMBER 2001). EXCERPTED TABLES

The referenced tables in their entirety are available on NTIA's website at http://www.ntia.doc.gov/reports/2008/NetworkedNation.html.

Table A-1. Household's Internet Connection Type, by selected characteristics of reference person: Total, Urban, Rural, and Central City, 2001 (Numbers in Thousands) Total USA

	Mode of internet access							
	Regular 'dial up' Telephone line		Digital Subscriber Line (DSL)		Cable modem		Other	
Total households with Internet access.	No.	%	No.	%	No.	%	No.	%
54,624	44,229	80.97	3,335	6.11	6,594	12.07	466	0.85

Table A-2. Household's Internet Connection Type, by selected characteristics of reference person: Total, Urban, Rural, and Central City, 2001 (Numbers in Thousands) Rural

Total households with Internet access.	Mode of internet access							
	Regular 'dial Up' Telephone line		Digital Subscriber Line (dsl)		Cable modem		Other	
	No.	%	No.	%	No.	%	No.	%
13,009	11,420	87.79	339	2.61	1,151	8.85	98	0.75

Table A-3. Households with Internet, by selected characteristics of reference person: Total, Urban, Rural, and CentraL City, 2001 (Numbers in Thousands) Total USA

Total hhlds	Households with internet											
	Yes										No	
	Thru computer		Thru cell phone/pager		Thru pda/other handheld		Thru tv-based internet		Thru ot device			
	No.	%	No.	%	No.	%	No.	%	No.	%	No.	%
108,562	53,783	49.54	5,139	4.73	1,893	1.74	700	0.65	256	0.24	53,938	49.68

Table A-4. Households with Internet, by selected characteristics of reference person: Total, Urban, Rural, and CentraL City, 2001 (Numbers in Thousands) Rural

Total HHLDS	Households with internet											
	Yes										No	
	Thru computer		Thru cell phone/pager		Thru pda/other handheld		Thru tv-based internet		Thru ot device			
	No.	%	No.	%	No.	%	No.	%	No.	%	No.	%
26,832	12,823	47.79	819	3.05	281	1.05	150	0.56	58	0.22	13,823	51.25

APPENDIX B: INTERNET USE SUPPLEMENT TO THE OCTOBER 2007. CURRENT POPULATION SURVEY U.S. CENSUS BUREAU (OCTOBER 2007). EXCERPTED TABLES

The referenced tables in their entirety are available on NTIA's website at http://www.ntia.doc.gov/reports/2008/NetworkedNation.html.

Table B-1. Households using the Internet in and outside the home, by selected characteristics: Total, Urban, Rural, Principal City, 2007 (Numbers in Thousands) Total USA

<table>
<tr><td rowspan="2"></td><td colspan="8">Households Using Internet</td><td colspan="2" rowspan="3">No internet use</td></tr>
<tr><td colspan="6">In the home</td><td colspan="2" rowspan="2">Anywhere</td></tr>
<tr><td rowspan="2">Total HHLDS</td><td colspan="2">Total</td><td colspan="2">Dial-up</td><td colspan="2">Broadband</td></tr>
<tr><td>No.</td><td>%</td><td>No.</td><td>%</td><td>No.</td><td>%</td><td>No.</td><td>%</td><td>No.</td><td>%</td></tr>
<tr><td>117,840</td><td>72,721</td><td>61.71</td><td>12,575</td><td>10.67</td><td>59,847</td><td>50.79</td><td>83,708</td><td>71.04</td><td>34,132</td><td>28.96</td></tr>
</table>

Table B-2. Households using the Internet in and outside the home, by selected characteristics: Total, Urban, Rural, Principal City, 2007 (Numbers in Thousands) Rural

	Households Using Internet								No Internet Use	
	In the Home						Anywhere			
Total HHLDS	Total		Dial-up		Broadband					
	No.	%	No.	%	No.	%	No.	%	No.	%
23,394	13,646	58.33	4,514	19.29	9,072	38.78	15,963	68.24	7,431	31.76

Table B-3. Percent of Households with Internet Broadband Access, Ranked by State: 2007 *(Numbers in Thousands)*

State	Total Households	Percent with Internet Broadband Access	90% Confidence Interval
New Hampshire	532	64.9	2.82
Alaska	244	62.5	2.95
Massachusetts	2,578	61.1	2.27
Connecticut	1,403	59.7	2.80
Utah	897	59.3	2.70
Rhode Island	420	59.2	2.94
Washington	2,656	58.4	2.30
Colorado	1,959	58.0	2.67
Hawaii	423	57.6	3.00
Oregon	1,513	57.5	2.85
New Jersey	3,292	57.1	2.07
California	13,037	56.4	1.04
Maryland	2,173	56.1	2.57
Kansas	1,115	55.2	2.87
Nevada	1,001	54.3	2.85
New York	7,471	54.1	1.39
Nebraska	713	54.1	2.87
Arizona	2,459	53.9	2.40
Georgia	3,663	53.9	1.95
Virginia	3,013	53.3	2.16
Florida	7,599	53.2	1.35
Minnesota	2,113	53.0	2.57
Wisconsin	2,394	52.6	2.43

Washington, DC	276	52.0	2.86
Illinois	4,980	51.6	1.69
Wyoming	215	50.4	2.97
Delaware	324	50.4	2.99
Ohio	4,611	48.8	1.76
North Dakota	274	48.7	2.79
Maine	541	48.4	3.12
Pennsylvania	5,056	47.7	1.68
Texas	8,794	47.6	1.28
South Dakota	324	47.5	2.64
North Carolina	3,548	47.1	2.00
Iowa	1,230	46.8	2.84
Vermont	258	46.8	3.04
Michigan	3,987	45.9	1.88
Idaho	569	45.5	2.77
Missouri	2,394	45.3	2.45
New Mexico	806	43.2	2.80
Louisiana	1,622	42.9	2.88
Indiana	2,579	42.3	2.30
Tennessee	2,492	41.6	2.34
Montana	424	40.2	2.67
Kentucky	1,749	40.0	2.77
South Carolina	1,783	39.1	2.73
Oklahoma	1,436	38.7	2.80
Arkansas	1,142	38.2	2.70
Alabama	1,866	37.4	2.62
Mississippi	1,137	33.2	2.68
West Virginia	752	32.7	2.46

INDEX

C

D

E

F

Q

R

S

T